HIGH-ORDER HARMONIC GENERATION IN
LASER PLASMA PLUMES

HIGH-ORDER HARMONIC GENERATION IN LASER PLASMA PLUMES

Rashid A. Ganeev
Imperial College London, UK

Imperial College Press

ICP

Published by

Imperial College Press
57 Shelton Street
Covent Garden
London WC2H 9HE

Distributed by

World Scientific Publishing Co. Pte. Ltd.
5 Toh Tuck Link, Singapore 596224
USA office: 27 Warren Street, Suite 401-402, Hackensack, NJ 07601
UK office: 57 Shelton Street, Covent Garden, London WC2H 9HE

British Library Cataloguing-in-Publication Data
A catalogue record for this book is available from the British Library.

HIGH-ORDER HARMONIC GENERATION IN LASER PLASMA PLUMES

ISBN 978-1-84816-980-7

Typeset by Stallion Press
Email: enquiries@stallionpress.com

Printed in Singapore by Mainland Press Pte Ltd.

To my parents, wife, son, and daughter

Preface

The application of laser plasma began in the early nineties with the aim of optimizing high-order harmonic sources of laser radiation in the extreme ultraviolet range. Unfortunately, the first experiments on harmonic generation in the passage of laser radiation through a plasma produced during laser ablation of a solid target turned out to be much less successful compared with harmonic generation in gaseous media. Those investigations stopped at the demonstration of relatively low-order harmonics and low efficiencies, which led to a decrease of interest in the "plasma harmonic" approach.

A new stage of studies started in 2005, when considerable improvements in harmonic generation in laser-produced plasmas led to revitalization of interest in this field of nonlinear optics. Here I show the developments of this new technique, which is currently widely applied for generation of coherent extreme ultraviolet radiation, as well as for harmonic generation for studies of matter.

This book represents the first comprehensive treatment of the subject, covering the principles, past and present experimental status, and important applications of high-order harmonic generation in laser-produced plasma plumes. I show how this method of frequency conversion of laser radiation towards the extreme ultraviolet range matured during multiple sets of studies carried out in many laboratories worldwide, and how it demonstrated new approaches in the generation of strong coherent short-wavelength radiation for various applications.

The book is based generally on studies carried out at the Institute for Solid State Physics, University of Tokyo (Kashiwa, Japan), Raja Ramanna Centre for Advanced Technology (Indore, India), Institut National de la Recherche Scientifique (Varennes, Canada), and Imperial College London (London, United Kingdom). I thank H. Kuroda, P.D. Gupta, P.A. Naik, T. Ozaki, J.P. Marangos, and J.W.G. Tisch for fruitful discussions and support at various stages of these studies.

I would like to thank my colleagues from the Institute of Electronics (Tashkent, Uzbekistan), where the experiments on plasma harmonics started. In particular, I would like to emphasize the role of T. Usmanov, I.A. Kulagin, V.I. Redkorechev, V.V. Gorbushin, R.I. Tugushev, G.S. Boltaev, and N.K. Satlikov, who were at the beginning of these studies and currently continue the high-order harmonic generation experiments in plasmas.

The important component of this research is a fruitful collaboration with many people involved in these studies outside the above-mentioned laser centers. I have enjoyed discussions with M.B. Danailov (ELETTRA, Trieste, Italy), B.A. Zon, M.V. Frolov, and N.L. Manakov (Voronezh State University, Russia), H. Zacharias (Westfälische Wilhelms-Universität, Münster, Germany), M. Castillejo (Instituto de Química Física Rocasolano, Madrid, Spain), D.B. Milošević (University of Sarajevo, Sarajevo, Bosnia and Herzegovina), J. Costello (Dublin City University, Dublin, Ireland), M. Lein and M. Tudorovskaya (Leibniz Universität Hannover, Hannover, Germany), E. Fiordilino (University of Palermo, Palermo, Italy), V. Strelkov (General Physics Institute, Moscow, Russia), M.K. Kodirov and P.V. Redkin (Samarkand State University, Samarkand, Uzbekistan), A.V. Andreev and S.Y. Stremoukhov (Moscow State University, Moscow, Russia), and C. Vozzi (Politecnico di Milano, Milan, Italy) regarding the past, present, and future joint experimental and theoretical studies of plasma harmonics.

I am also indebted to the members of those and other teams for collaboration during this research. In particular, I would like to emphasize the contributions of M. Suzuki, M. Baba, L.B. Elouga Bom, H. Singhal, J.A. Chakera, U. Chakravarty, R.A. Khan, M. Raghuramaiah, V. Arora, S.R. Kumbhare, R.P. Kushwaha, M. Tayyab, V.R. Bhardvaj, C. Hutchison, T. Witting, F. Frank, W.A. Okell, T. Siegel, A. Zaïr, S. Weber, I. Sola, M.E. Lopez-Arias, M. Oujja, M. Sanz, I. López-Quintás, and M. Martin during the studies of harmonic generation in laser-produced plasmas.

Finally, I would like to thank my wife Lida, son Timur, and daughter Dina for their patience during my long trips worldwide and support at all stages of my life.

Rashid Ganeev

Contents

Contents | xiii

List of Figures and Tables

Chapter 1

Chapter 2

Chapter 3

Chapter 4

Chapter 5

Chapter 6

Chapter 7

1

Introduction

The quest for efficient coherent short-wavelength radiation sources for use in biology, plasma diagnostics, medicine, microscopy, photolithography, etc., has a long history. For a long time, three main approaches were considered for these purposes, namely X-ray lasers [1–3], free-electron lasers [4, 5], and high-order harmonic generation (HHG) of laser radiation by different means [6, 7]. Though the Web of Science search database gives approximately equal numbers of publications on those topics, the third approach seems very attractive from the point of view of the availability of moderate energy lasers in many laboratories worldwide, and lower expenditure for everyday use compared with the two former methods. Other disadvantages of X-ray lasers are their poor spatial coherence and radiation divergence. As regards free-electron lasers that generate radiation in the short-wavelength spectral range, there are only a limited number of sources reported so far. Furthermore, the application of these lasers is largely limited by their high cost.

Currently, HHG research is actively being pursued due to the availability of new high-power compact near infrared (IR) laser systems offering excellent output parameters (high energy and intensity of pulses and high pulse repetition rate). Two mechanisms are used for HHG: harmonic generation in gases and from surfaces. The considerable progress achieved in this area has enabled extending the range of generated coherent radiation to the spectral region where the radiation can pass through water-bearing components (the so-called water window, 2.3–4.6 nm), which is extremely important for various medical and biological applications.

1

In isotropic media, HHG using moderate-level femtosecond laser pulses allows easy production of coherent radiation in the extreme ultraviolet (XUV) spectral range. During the last 25 years, predominantly rare gases were employed as target media for HHG, which, however, imposed some physical and practical limits on the performance of these coherent XUV sources. So far, only low conversion efficiencies of HHG have been reported using gases as the nonlinear media, despite enormous efforts. For practical applications of high-order harmonic sources, higher conversion efficiency and thus an increase in the photon flux and also of the maximum photon energy of the harmonic radiation would be beneficial. The generation of high-order harmonics in another isotropic medium, laser-produced plasma, being for this purpose a relatively new and largely unexplored medium, promises to yield these expectations, as well as to open the door for new developments in laser–matter interactions.

In this book, we do not consider HHG in gases, because this topic has been covered by recent comprehensive papers (see for example the review [8] and references therein). Nor do we consider the mechanisms of odd and even harmonics generation in the reflection of laser radiation from surfaces, because this problem is discussed at length in several reviews and papers ([9,10] and references therein) and, furthermore, is beyond the scope of our consideration. Our aim is to familiarize the reader with the most recent approaches to harmonic generation in the XUV range with the use of an isotropic medium (different from that used previously in gas-jet sources and special gas-filled cells). We primarily discuss the experimental and theoretical results recently reported in this field.

The application of laser plasma came up at the beginning of the nineties with the aim of optimizing HHG light sources. The idea was based on the use of the larger ionization potentials of alkali ions as compared to noble gases. The use of such plasma was aimed at improving the HHG phase matching conditions, and increasing the concentration of excited ions and neutrals in plasma, which was thought to be the right way for amendments of harmonic yield. In the meantime, the first experiments on HHG in the passage of laser radiation through a plasma produced during laser ablation of a solid target turned out to be much less successful. Data obtained with the use of highly excited plasmas containing multiply charged ions revealed several limiting factors, which did not permit generation of harmonics of sufficiently high orders [11–16]. Those investigations stopped after the demonstration of

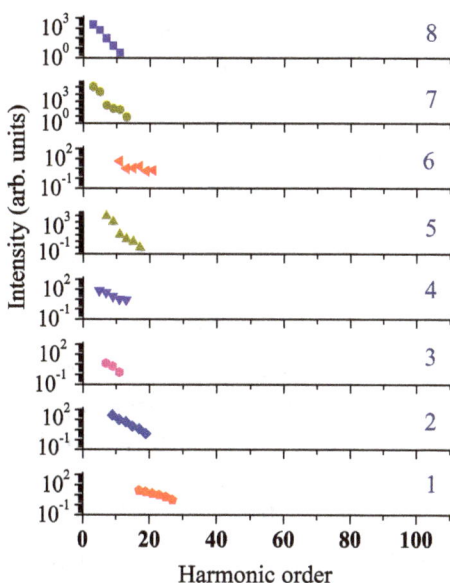

Fig. 1.1 High-order harmonic spectra obtained from laser-produced plasma during the early stages of this research (1992–1997). 1: K II [13], 2: LiF II [16], 3: Na II [11], 4: C II [14], 5: C II [12], 6: Pb III [12], 7: Li II [11], 8: Al II [15]. Adapted from [17] with permission from John Wiley & Sons.

relatively low-order harmonics (Fig. 1.1 [17]). These disadvantages, as well as low conversion efficiency (10^{-7}–10^{-6}), led to the erosion of interest in this HHG technique, especially in comparison with the achievements involving gas HHG sources.

However, there were reasons to hope that harmonic intensities could be increased and efficient shorter-wavelength coherent radiation sources could be developed using laser-produced plasmas. There are no fundamental limitations here; it only remained to find the optimal conditions for producing an appropriate plasma plume to serve as the efficient nonlinear medium for HHG. Laser-produced plasma may be validly used for this process if the effect of the limiting factors (self-defocusing and the wave phase mismatch of the harmonics and the radiation being converted) is minimized. The authors of Ref. [11] noted that, by fulfilling these conditions, the plasma produced on the surface of the target should be an efficient medium for HHG, whatever the target material is. Anticipating developments, we note that recent HHG research has exhibited the decisive effect of the physical parameters of the targets on the properties of the frequency conversion of laser radiation.

The advantages of plasma HHG could largely be realized with the use of low-excited and weakly ionized plasma, because the limiting processes governing the dynamics of laser frequency conversion would play a minor role in this case. Attention was drawn to this feature early in the study of third-harmonic generation in weakly ionized plasma [18]. This assumption allowed the formulation of several recommendations for further advancement toward the development of efficient shorter wavelength sources based on the frequency conversion of laser radiation in various plasmas. A new history of plasma HHG studies based on this approach started in 2005 [19]. It was immediately followed by observation of extended harmonics and considerably higher conversion efficiencies ($>10^{-5}$ [20–22]), which became comparable with those reported in gas HHG studies. During the following years, enormous improvements in the characteristics of this process were reported. A substantial increase in the highest order of the generated harmonics, emergence of a plateau in the energy distribution of highest-order harmonics, high efficiencies obtained with several plasma formations, realization of resonance-induced enhancement of individual harmonics, efficient harmonic enhancement for plasma plumes containing clusters of different materials, and other properties revealed in these and other works [23–33] have demonstrated the advantages of using specially prepared plasmas for HHG.

Nowadays, the orders of harmonics generated from plasma media can be easily extended into the sixties and seventies. The highest-order harmonics (the 101st harmonic) have been obtained in manganese plasma [34]. The harmonic efficiency in metal plasmas in the plateau region amounts to $\sim 10^{-5}$ [35, 36]. In addition, the efficiency of conversion to an individual (resonantly enhanced) high-order harmonic has approached 10^{-4} [37]. Recent studies of carbon plasma have revealed that the high-order harmonics from this medium can also achieve 10^{-4} conversion efficiency, thus reaching the microjoule level of pulse energies for these XUV sources [38, 39]. The majority of these parameters are at the same level as the results obtained in gas media. But the highest orders of harmonics produced by these two techniques still remain different. For gases, the generation of harmonics of orders exceeding 2000 has been reported [40]. The substantial difference in the orders of harmonics generated in gases and in

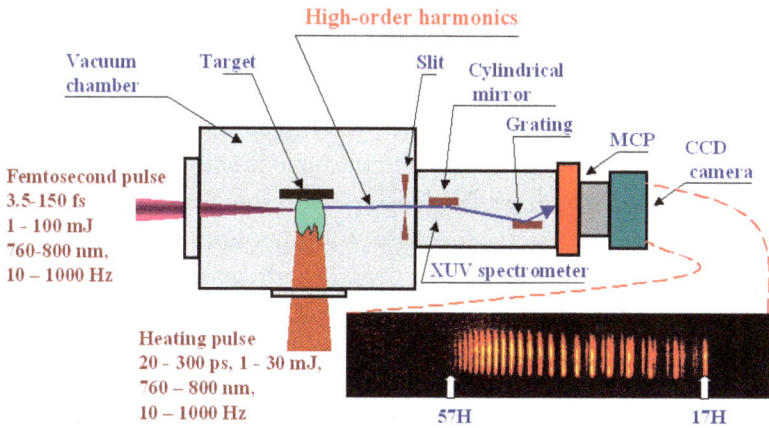

Fig. 1.2 A schematic view illustrating high-order harmonic generation in laser-produced plasma. Adapted from [17] with permission from John Wiley & Sons.

plasmas raises the problem of searching for new ways to produce higher-energy coherent photons by means of HHG in plasma plumes. This opportunity can be provided by doubly charged ions, which increase the maximum attainable energy of the generated coherent short-wavelength photons.

The experimental setups commonly used in the studies of HHG from plasmas are similar to the one presented in Fig. 1.2 [17]. In reported studies, the parameters of both laser and plasma were varied in a broad range, while the optimization of harmonic yield was carried out to achieve the highest conversion efficiency and extension of harmonics toward the XUV range. Briefly, to create the ablation, a heating pulse that was split from the uncompressed radiation of a Ti:sapphire laser (pulse duration $t \approx 10$–300 ps, wavelength $\lambda \approx 760$–800 nm, pulse energy $E \approx 1$–30 mJ, pulse repetition rate 10 Hz – 1 kHz) was focused on a target placed in the vacuum chamber to create a plasma plume. After some delay (of the order of few tens nanoseconds), the compressed femtosecond pulse ($E = 1$–100 mJ, $t = 4$–150 fs, $\lambda = 760$–800 nm) was focused onto the plasma from the orthogonal direction. The generating harmonics were spectrally dispersed by XUV spectrometers. The XUV spectrum was then detected by a micro-channel plate (MCP) and finally recorded using a charge-coupled device (CCD). A description of the details of plasma harmonic experiments can be found in the above-mentioned references [30, 32, 34].

Recent developments of this technique have revealed many new approaches and achievements in harmonic generation from plasmas. Among them are the studies of harmonics from clusters using the ablation of commercially available nanoparticles [41, 42], fullerenes [43–47], carbon nanotubes [48], resonance-enhanced features of odd and even harmonics [49], generation of extremely broadband high-order harmonics [50], application of high pulse repetition rate laser sources and ultrashort pulses for HHG in plasma plumes [31], observation of quantum path signatures in the harmonic spectra from various plasmas [32], enhancement of harmonics from *in-situ*-produced clusters [51], development of theoretical approaches describing the observed peculiarities of resonance-enhanced harmonics [52–57], emergence of a "second" plateau in harmonic distribution [58], development of two-color pump schemes for plasma-induced harmonics [59, 60], observation of extremely strong HHG in carbon-contained plasmas [38, 39], proposals for quasi-phase matching in plasma plumes [61], observation of the attosecond nature of pulse duration of plasma-induced harmonics [62], etc. All of these findings will be discussed in this book. Among future developments in the application of this technique, one may anticipate such areas as the seeding of plasma resonance harmonics in XUV free-electron lasers, application of endohedral fullerenes for plasma HHG, analysis of molecular structures through the study of harmonic spectra from oriented molecules in plasmas, the search for quasi-phase-matching schemes in plasma plumes, generation of strong combs and single attosecond pulses, application of IR (1000–3000 nm) laser sources for extension of plasma harmonic cutoffs, to mention a few.

Recent reviews on plasma harmonics mostly concentrated on discussion of such specific topics as application of nanoparticle-containing plasmas for HHG [63], resonance-induced enhancement of harmonics [64], and applications of fullerenes as the attractive media for harmonic generation [65], in contrast to the two first topical reviews [36, 66], where the whole spectrum of plasma harmonics studies was presented. The basic ideas of HHG in laser-produced plasmas were outlined in previous reviews. It seems timely to return to the practice of showing the broad pattern of various developments in this field [17, 67]. It is also obvious that a comprehensive overview of recent findings can help in defining the next steps of development in this relatively new and attractive area of nonlinear optical studies. In the following sections, we will describe the most important findings and highlight some new work in this field as well as offer a detailed overview of related studies. In addition we give an outline of the expectations of the future evolution of HHG studies in laser-produced plasma plumes.

The structure of this book is as follows. The basic principles of harmonic generation in laser-produced plasmas are described in Chapter 2. Resonance-induced enhancement of HHG is discussed in Chapter 3. The application of cluster-containing plasma plumes as efficient media for harmonic generation is described in Chapter 4. Chapter 5 shows the studies of fullerenes as the media for HHG. The new methods of enhancement of the harmonic yield from plasma plumes are discussed in Chapter 6. Finally, new trends in plasma HHG are presented in Chapter 7, together with a summary of the current status of HHG in laser-produced plasmas and discussion of the future perspectives in this field.

References

1. Desai, T., Daido, H., Suzuki, M. *et al.* (2011). X-ray emission from laser-irradiated gold targets with surface modulation, *Laser Part. Beams*, **19**, 241–247.
2. Ozaki, T., Ganeev, R.A., Ishizawa, A. *et al.* (2002). Highly directive 18.9 nm nickel-like molybdenum x-ray laser operating at 150 mJ pump energy, *Phys. Rev. Lett.*, **89**, 253902.
3. Suckewer, S. and Jaegle, P. (2009). X-ray laser: Past, present, and future, *Laser Phys. Lett.*, **6**, 411–436.
4. Patterson, B.D. and Abela, R. (2010). Novel opportunities for time-resolved absorption spectroscopy at the X-ray free electron laser, *Phys. Chem. Chem. Phys.*, **12**, 5647–5652.
5. Kur, E., Dunning, D.J., McNeil, B.W.J. *et al.* (2011). A wide bandwidth free-electron laser with mode locking using current modulation, *New J. Phys.*, **13**, 063012.
6. Norreys, P.A., Zepf, M., Moustaizis, S. *et al.* (1996). Efficient extreme UV harmonics generated from picosecond laser pulse interactions with solid targets, *Phys. Rev. Lett.*, **76**, 1832–1835.
7. Vozzi, C., Calegari, F., Frassetto, F. *et al.* (2010). High order harmonics driven by a self-phase-stabilized IR parametric source, *Laser Phys.*, **20**, 1019–1027.
8. Zhang, G.P. (2007). High harmonic generation in atoms, molecules and nanostructures, *Int. J. Mod. Phys. B*, **21**, 5167–5185.
9. Gibbon, P. (2005). *Short Pulse Laser Interactions with Matter: An Introduction*, Imperial College Press, London.
10. Teubner, U. and Gibbon, P. (2009). High-order harmonics from laser-irradiated plasma surfaces, *Rev. Mod. Phys.*, **81**, 445–479.
11. Akiyama, Y., Midorikawa, K., Matsunawa, Y. *et al.* (1992). Generation of high-order harmonics using laser-produced rare-gas-like ions, *Phys. Rev. Lett.*, **69**, 2176–2179.
12. Kubodera, S., Nagata, Y., Akiyama, Y. *et al.* (1993). High-order harmonic generation in laser-produced ions, *Phys. Rev. A*, **48**, 4576–4582.

13. Wahlström, C.-G., Borgström, S., Larsson, J. *et al.* (1995). High-order harmonic generation in laser-produced ions using a near-infrared laser, *Phys. Rev. A*, **51**, 585–591.
14. Theobald, W., Wülker, C., Schäfer, F.R. *et al.* (1995). High-order harmonic generation in carbon vapor and low charged plasma, *Opt. Commun.*, **120**, 177–183.
15. Ganeev, R.A., Redkorechev, V.I., and Usmanov, T. (1997). Optical harmonics generation in low-temperature laser produced plasma, *Opt. Commun.*, **135**, 251–256.
16. Krushelnick, K., Tighe, W., and Suckewer, S. (1997). Harmonic generation from ions in underdense aluminum and lithium–fluorine plasmas, *J. Opt. Soc. Am. B*, **14**, 1687–1691.
17. Ganeev, R.A. (2012). Generation of harmonics of laser radiation in plasmas, *Laser Phys. Lett.*, **9**, 175–194.
18. Gladkov, S.M. and Koroteev, N.I. (1990). Quasiresonant nonlinear optical processes involving excited and ionized atoms, *Sov. Phys. Usp.*, **33**, 554–575.
19. Ganeev, R., Suzuki, M., Baba, M. *et al.* (2005). High-order harmonic generation from boron plasma in the extreme-ultraviolet range, *Opt. Lett.*, **30**, 768–770.
20. Ganeev, R.A., Singhal, H., Naik, P.A. *et al.* (2006). Harmonic generation from indium-rich plasmas, *Phys. Rev. A*, **74**, 063824.
21. Ganeev, R.A., Baba, M., Suzuki, M. *et al.* (2006). Optimization of harmonic generation from boron plasma, *J. Appl. Phys.*, **99**, 103303.
22. Ganeev, R.A., Suzuki, M., Baba, M. *et al.* (2005). Generation of strong coherent extreme ultraviolet radiation from the laser plasma produced on the surface of solid targets, *Appl. Phys. B*, **81**, 1081–1089.
23. Ganeev, R.A., Singhal, H., Naik, P.A. *et al.* (2006). Single harmonic enhancement by controlling the chirp of the driving laser pulse during high-order harmonic generation from GaAs plasma, *J. Opt. Soc. Am. B*, **23**, 2535–2540.
24. Suzuki, M., Baba, M., Ganeev, R. *et al.* (2006). Anomalous enhancement of single high-order harmonic using laser ablation tin plume at 47 nm, *Opt. Lett.*, **31**, 3306–3308.
25. Suzuki, M., Baba, M., Kuroda, H. *et al.* (2007). Intense exact resonance enhancement of single-high-harmonic from an antimony ion by using Ti:sapphire laser at 37 nm, *Opt. Express*, **15**, 1161–1166.
26. Ganeev, R.A., Naik, P.A., Singhal, H. *et al.* (2007). Strong enhancement and extinction of single harmonic intensity in the mid- and end-plateau regions of the high harmonics generated in low-excited laser plasmas, *Opt. Lett.*, **32**, 65–67.
27. Elouga Bom, L.B., Kieffer, J.-C., Ganeev, R.A. *et al.* (2007). Influence of the main pulse and prepulse intensity on high-order harmonic generation in silver plasma ablation, *Phys. Rev. A*, **75**, 033804.
28. Ozaki, T., Elouga Bom, L.B., Ganeev, R.A. *et al.* (2008). Extending the capabilities of ablation harmonics to shorter wavelengths and higher intensity, *Laser Part. Beams*, **26**, 235–240.
29. Ganeev, R.A., Naik, P.A., Chakera, J.A. *et al.* (2011). Carbon aerogel plumes as an efficient medium for higher harmonic generation in 40–90 nm range, *J. Opt. Soc. Am. B*, **28**, 360–364.
30. Ganeev, R.A., Suzuki, M., Baba, M. *et al.* (2007). High harmonic generation from the laser plasma produced by the pulses of different duration, *Phys. Rev. A*, **76**, 023805.
31. Ganeev, R.A., Hutchison, C., Siegel, T. *et al.* (2011). High-order harmonic generation from metal plasmas using 1 kHz laser pulses, *J. Mod. Opt.*, **58**, 819–824.

32. Ganeev, R.A., Hutchison, C., Siegel, T. *et al.* (2011). Quantum path signatures in harmonic spectra from metal plasma, *Phys. Rev. A*, **83**, 063837.
33. Ganeev, R.A., Elouga Bom, L.B. and Ozaki, T. (2011). Time-resolved spectroscopy of plasma plumes: a versatile approach for optimization of high-order harmonic generation in laser plasma, *Phys. Plasmas*, **18**, 083101.
34. Ganeev, R.A., Elouga Bom, L.B., Kieffer, J.-C. *et al.* (2007). Demonstration of the 101st harmonic generated from laser-produced manganese plasma, *Phys. Rev. A*, **76**, 023831.
35. Ganeev, R.A., Baba, M., Suzuki, M. *et al.* (2005). High-order harmonic generation from silver plasma, *Phys. Lett. A*, **339**, 103–109.
36. Ganeev, R.A. (2009). Generation of high-order harmonics of high-power lasers in plasmas produced under irradiation of solid target surfaces by a prepulse, *Phys. Usp.*, **52**, 55–77.
37. Ganeev, R.A., Suzuki, M., Ozaki, T. *et al.* (2006). Strong resonance enhancement of a single harmonic generated in extreme ultraviolet range, *Opt. Lett.*, **31**, 1699–1701.
38. Elouga Bom, L.B., Petrot, Y., Bhardwaj, V.R. *et al.* (2011). Multi-μJ coherent extreme ultraviolet source generated from carbon using the plasma harmonic method, *Opt. Express*, **19**, 3077–3085.
39. Petrot, Y., Elouga Bom, L.B., Bhardwaj, V.R. *et al.* (2011). Pencil lead plasma for generating multimicrojoule high-order harmonics with a broad spectrum, *Appl. Phys. Lett.*, **98**, 101104.
40. Seres, E., Seres, J., and Spielmann, C. (2006). X-ray absorption spectroscopy in the keV range with laser generated high harmonic radiation, *Appl. Phys. Lett.*, **89**, 181919.
41. Ganeev, R.A., Elouga Bom, L.B., and Ozaki, T. (2009). Application of nanoparticle-containing laser plasmas for harmonic generation, *J. Appl. Phys.*, **106**, 023104.
42. Ozaki, T., Elouga Bom, L.B., Abdul-Haji, J. *et al.* (2010). Evidence of strong contribution from neutral atoms in intense harmonic generation from nanoparticles, *Laser Part. Beams*, **28**, 69–74.
43. Ganeev, R.A., Elouga Bom, L.B., Abdul-Hadi, J. *et al.* (2009). High-order harmonic generation from fullerene using the plasma harmonic method, *Phys. Rev. Lett.*, **102**, 013903.
44. Ganeev, R.A., Elouga Bom, L.B., Wong, M.C.H. *et al.* (2009). High-order harmonic generation from C_{60}-rich plasma, *Phys. Rev. A*, **80**, 043808.
45. Ganeev, R.A., Singhal, H., Naik, P.A. *et al.* (2009). Influence of C_{60} morphology on high-order harmonic generation enhancement in fullerene-containing plasma, *J. Appl. Phys.*, **106**, 103103.
46. Ganeev, R.A., Singhal, H., Naik, P.A. *et al.* (2010). Enhanced harmonic generation in C_{60}-containing plasma plumes, *Appl. Phys. B*, **100**, 581–585.
47. Ganeev, R.A., Baba, M., Kuroda, H. *et al.* (2011). Low- and high-order nonlinear optical characterization of C_{60}-containing media, *Eur. Phys. J. D*, **64**, 109–114.
48. Ganeev, R.A., Naik, P.A., Singhal, H. *et al.* (2011). High order harmonic generation in carbon nanotube-containing plasma plumes, *Phys. Rev. A*, **83**, 013820.
49. Ganeev, R.A., Chakera, J.A., Naik, P.A. *et al.* (2011). Resonance enhancement of single even harmonic in tin-containing plasma using intensity variation of two-color pump, *J. Opt. Soc. Am. B*, **28**, 1055–1061.
50. Ganeev, R.A. and Kuroda, H. (2009). Extremely broadened high-order harmonics generated by the femtosecond pulses propagating through the filaments in air, *Appl. Phys. Lett.*, **95**, 201117.

51. Singhal, H., Ganeev, R.A., Naik, P.A. *et al.* (2010). In-situ laser induced silver nanoparticle formation and high order harmonic generation, *Phys. Rev. A*, **82**, 043821.

52. Kulagin, I.A. and Usmanov, T. (2009). Efficient selection of single high-order harmonic caused by atomic autoionizing state influence, *Opt. Lett.*, **34**, 2616–2619.

53. Strelkov, V. (2010). Role of autoionizing state in resonant high-order harmonic generation and attosecond pulse production, *Phys. Rev. Lett.*, **104**, 123901.

54. Milošević, D.B. (2010). Resonant high-order harmonic generation from plasma ablation: Laser intensity dependence of the harmonic intensity and phase, *Phys. Rev. A*, **81**, 023802.

55. Frolov, M.V., Manakov, N.L., and Starace, A.F. (2010). Potential barrier effects in high-order harmonic generation by transition-metal ions, *Phys. Rev. A*, **82**, 023424.

56. Redkin, P.V. and Ganeev, R.A. (2010). Simulation of resonant high-order harmonic generation in three-dimensional fullerenelike system by means of multiconfigurational time-dependent Hartree–Fock approach, *Phys. Rev. A*, **81**, 063825.

57. Tudorovskaya, M. and Lein, M. (2011). High-order harmonic generation in the presence of a resonance, *Phys. Rev. A*, **84**, 013430.

58. Ganeev, R.A., Suzuki, M., Baba, M. *et al.* (2009). Extended high-order harmonic spectra from the laser-produced Cd and Cr plasmas, *Appl. Phys. Lett.*, **94**, 051101.

59. Ganeev, R.A., Singhal, H., Naik, P.A. *et al.* (2009). Enhancement of high-order harmonic generation using two-color pump in plasma plumes, *Phys. Rev. A*, **80**, 033845.

60. Ganeev, R.A., Singhal, H., Naik, P.A. *et al.* (2010). Systematic studies of two-color pump induced high order harmonic generation in plasma plumes, *Phys. Rev. A*, **82**, 053831.

61. Sheinfux, A.H., Henis, Z., Levin, M. *et al.* (2011). Plasma structures for quasiphase matched high harmonic generation, *Appl. Phys. Lett.*, **98**, 141110.

62. Elouga Bom, L.B., Haessler, S., Gobert, O. *et al.* (2011). Attosecond emission from chromium plasma, *Opt. Express*, **19**, 3677–3685.

63. Ganeev, R.A. (2008). High-order harmonic generation in nanoparticle-containing laser-produced plasmas, *Laser Phys.*, **18**, 1009–1015.

64. Ganeev, R.A. (2009). Application of resonance-induced processes for enhancement of the high-order harmonic generation in plasma, *Open Spectrosc. J.*, **3**, 1–8.

65. Ganeev, R.A. (2011). Fullerenes: The attractive medium for harmonic generation, *Laser Phys.*, **21**, 25–43.

66. Ganeev, R.A. (2007). High-order harmonic generation in laser plasma: A review of recent achievements, *J. Phys. B*, **40**, R213–R253.

67. Ganeev, R.A. (2012). Harmonic generation in laser-produced plasma containing atoms, ions and clusters: A review, *J. Mod. Opt.*, **59**, 409–439.

2

Basic Principles of Harmonic Generation in Plasmas

In this chapter, we introduce the fundamentals of HHG in isotropic media, present some examples of experiments on HHG in various laser plasmas, show the advantages of using shorter wavelengths for harmonic generation in laser plasmas, compare HHG in plasmas produced by laser pulses of various durations, and analyze laser-produced plasma characteristics for optimization of HHG.

2.1. Fundamentals of HHG in Isotropic Media

As already mentioned, the high-order harmonics of laser radiation may be generated by three different techniques: (i) in the interaction of laser pulses (radiation intensity $I \approx 10^{14}$–10^{16} W cm^{-2}) with gas jets or a specially prepared gas located in cells and waveguides; (ii) in the interaction of laser pulses with a higher intensity ($I > 10^{18}$ W cm^{-2}) and a high contrast ratio (10^7 and higher) with the surfaces of solids; and (iii) in the passage of laser radiation ($I \approx 10^{14}$–10^{15} W cm^{-2}) through specially prepared plasma media. We do not consider the second technique here. Investigations of HHG from surfaces are analyzed in detail in monograph [1] and review [2]. We only briefly touch upon the problems arising in HHG from gases. These processes have been analyzed explicitly in monograph [3]. High-order harmonic generation in gases has also been investigated in comprehensive reviews [4, 5]. Our main topic is the analysis of HHG in the passage of moderately intense radiation through laser-produced plasmas, i.e., the plasmas produced on the surfaces of different solid targets using additional subnanosecond laser pulses timed with the femtosecond laser pulse.

When a high-intensity laser pulse passes through a gaseous or plasma medium, its atoms and ions emit odd harmonics. For a laser radiation wavelength λ, a superposition of the components λ, $\lambda = 3$, $\lambda = 5$, $\lambda = 7$, etc. is observed at the output of the nonlinear medium. The harmonics of laser radiation result from a three-stage process [6–8], which is described below. This process is periodically repeated every half cycle of the electromagnetic wave. The highest-order harmonics are due to the electron acceleration at the instant of ionization at the peak intensity of the laser pulse. Therefore, the generation of highest-order harmonics results from the interaction of a high-intensity light field with atoms [9–11], atomic clusters [12, 13], molecules [14, 15], and ions [16–29].

Specifically, the process of HHG in atoms and ions can be described in the following way (Fig. 2.1a). When an intense laser pulse interacts with an atom or ion, an electron may be ejected. However, for linearly polarized fields of laser radiation and at certain times t the electron can return to the atom during the next half-cycle of the laser field.

As a result of possible recombination of the electron with the atom, high-frequency photons are emitted. The resulting spectrum of radiation (HHG power spectrum) has the following typical features:

- Discrete nature. The power spectrum consists of high harmonics of the pump radiation with frequencies $N\omega_0$. For media with a center of inversion (namely, atomic plasmas), N is an odd number. In the case of symmetry breakdown (for example, by a strong magnetic field) or when using pump radiation with a contribution of its even harmonic (usually the second harmonic), a spectrum that consists of both odd and even harmonics is generated. The width of the peaks of harmonics depends on the length of the pulse as $\Delta\omega \approx 2\pi/T$.

- Existence of a plateau that is a frequency range with harmonics of approximately equal intensities. This forms a contrast with a strong decrease of the intensity with an increase of the harmonic order both for the lowest and for the highest harmonics produced.

- Existence of cutoff — a rapid decrease of intensity at the end of the plateau after the cutoff harmonic N_C. The cutoff harmonic can be determined

(a)

(b)

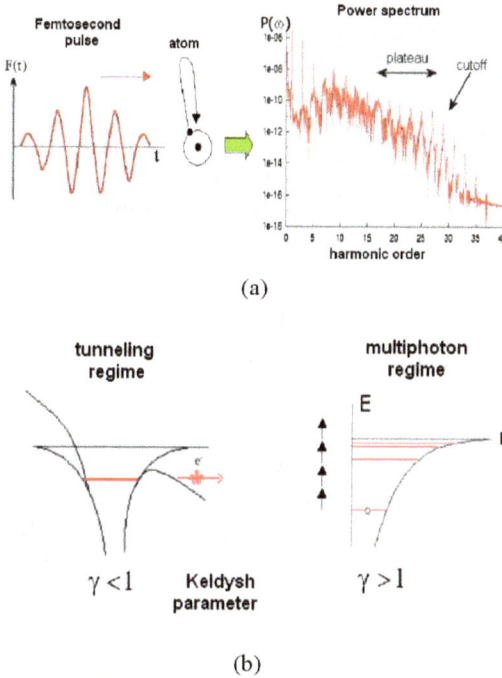

Fig. 2.1 (a) Principles of HHG. (b) Different regimes of ionization.

classically by the maximum energy that the electron accelerated by the laser field can have at the moment of recombination with the atom.

Some typical features of the power spectrum have a classical explanation, at least for high intensities of laser fields. For this purpose let us consider the behavior of a free classical electron under the influence of the field $F(t) = F_0 \sin(\omega_0 t)$. The field amplitude $F_0 = \sqrt{2I_A\sqrt{\varepsilon_0/\mu_0}}$ can be derived from its peak intensity in atomic units:

$$I_A \text{ (a.u.)} = 1.555 \times 10^{-16} I_A \text{ (W cm}^{-2}) \tag{2.1}$$

The movement of an electron follows the oscillations of the field with amplitude $\alpha = F_0/\omega_0^2$. From the maximal speed of an electron during this oscillation $v_{max} = F_0/\omega_0$, one can derive ponderomotive energy:

$$U_P = v_{max}^2/2 = F_0^2/(4\omega_0^2) = \frac{1}{2}\sqrt{\frac{\varepsilon_0}{\mu_0}} \cdot \frac{I_A}{\omega_0^2}. \tag{2.2}$$

The Keldysh parameter $\gamma = \sqrt{I_P/2U_P}$ is the qualitative estimate of relative importance of ponderomotive energy U_P and ionization energy I_P (Fig. 2.1b). For $\gamma > 1$, when U_P is small compared to the ionization energy, the electron is mainly ejected by absorption of multiple photons (multiphoton regime). But in the tunnel regime ($\gamma < 1$) Coulomb potential is greatly suppressed by the laser field, so the atom is mainly ionized by tunneling through the barrier.

For the description of HHG in the tunnel regime a semiclassical three-step model is often used. This model divides the HHG process into tunneling, classical movement, and recombination with ion.

- In the first step the electron is ejected by tunneling. The interaction with the electric field $F(t) = F_0 \sin(\omega_0 t)$ changes the atom's Coulomb potential to a Stark potential. This creates a tunnel barrier at $r_B(t) = \sqrt{1/F(t)}$.
- The second step is fully classical. The ejected electron is described classically and its trajectory under the influence of the laser field is determined. Coulomb attraction to the nucleus is not taken into account. For simplicity, the electron is considered to be born at the center of the atom with zero initial velocity. The tunneling time is reflected by the phase ϕ relative to the laser field. During the following return to the nucleus the corresponding to return phase ψ classical energy and return time are determined.
- The third step is stimulated transition of an electron from the state induced by the strong laser field to the ground state, which is accompanied by the emission of harmonics.

Let us determine the energy of cutoff harmonic N_C. The movement of an electron ejected at the time $t = 0$ relative to laser field $F(t)$ with phase ϕ is described by a system of equations:

$$a_\phi(t) = F_0 \sin(\omega t + \phi)$$

$$v_\phi(t) = \frac{F_0}{\omega_0}[\cos(\phi) - \cos(\omega t + \phi)] \tag{2.3}$$

$$x_\phi(t) = \frac{F_0}{\omega_0^2}[\omega t \cos(\phi) + \sin(\phi) - \sin(\omega t + \phi)]$$

The velocity v_ϕ is proportional to the area under the field starting from the phase ϕ. Due to symmetry, the turning point of an electron wavepacket lies at $\phi' = 2\pi - \phi$. The electron can return to the center only if ϕ is inside the interval $\phi = \lceil \pi/2, \pi \rceil$. We determine the phase which corresponds to the collision time t as $\psi = \omega t' + \phi$, then we get the relation between ϕ and ψ:

$$(\psi - \phi) \cos (\phi) + \sin (\phi) - \sin (\psi) = 0. \tag{2.4}$$

The kinetic energy $E_{kin} = v^2/2$ at the time of collision is

$$E_{kin} = \frac{1}{2} \left[\frac{F_0}{\omega} (\cos (\phi) - \cos (\psi)) \right]^2 \tag{2.5}$$

Taking into account (2.4), the maximal kinetic energy $E_{kinmax} = 3.17 U_P$ is observed at $\phi = 107°$ and $\psi = 345°$ for the trajectories with maximum possible energy. Because the electron wavepacket is influenced by the ion's field during the collision, ionization energy $I_P = |E_0|$ is being added. We get the following values for cutoff energy and cutoff harmonic:

$$E_C = I_P + 3.17 U_P \quad N_C = E_C/\omega_0 \tag{2.6}$$

The relation (2.6) is also useful because it helps to determine which kind of ions is responsible for the given cutoff harmonic. It should be noted that relation (2.6) is derived semiclassically and may be not conserved for relatively complex systems.

Along with the microscopic consideration of the processes occurring in the interaction of high-power ultrashort laser pulses with atoms and ions, account should also be taken of macroscopic processes such as the effect of transmission through a medium and group effects. These effects primarily include dephasing, absorption, and defocusing, and are analyzed in [30].

2.2. High-Order Harmonic Generation in Various Laser Plasmas

As noted in Chapter 1, one of the principal goals of HHG studies is the search for plasma media that allow maximization of the generated harmonic orders and an increase of the energy of short-wavelength pulses [31–33]. In this section, we analyze the frequency conversion of laser radiation in plasmas of different materials for the purpose of reaching this goal.

2.2.1. *Boron*

Two HHG schemes have been used in studies of laser radiation frequency conversion in boron plasmas. In the first (orthogonal) scheme, which was used in the majority of HHG investigations, part of the uncompressed Ti:sapphire laser radiation generated the plasma, following which femtosecond radiation was focused into the plasma from an orthogonal direction with a certain delay. In the second (longitudinal) scheme, a subnanosecond pulse was focused onto a surface region adjacent to a bore drilled in the target and the pulse produced the plasma, following which a second (femtosecond) pulse was directed parallel to the first one and passed through the plasma and the bore. The harmonics generated in this case were recorded using the technique described in Chapter 1.

High-order harmonics up to order 63 ($\lambda \approx 12.6$ nm [34, 35]) were observed in experiments with boron plasmas conducted following the orthogonal scheme. A plateau-like pattern showed up in the distribution of harmonics above the seventh order (Fig. 2.2) and the pattern was similar to the distribution of harmonics in numerous experiments with "gas" harmonics. The efficiency of conversion to harmonics varied from 10^{-4} (for the third order) to 10^{-7} for harmonics located in the plateau region. The plateau-like pattern vanished when the heating pulse intensity at the target exceeded some limit at which a large number of free electrons were generated and a continuum was observed

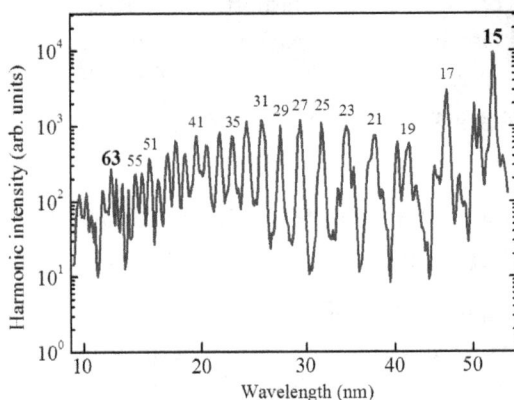

Fig. 2.2 Spectrum of the harmonics generated in boron plasma. Reprinted from [34] with permission from Optical Society of America.

in the plasma emission. The highest harmonic order did not then exceed 19. With a further increase in the heating pulse intensity at the target, intense lines of doubly charged ions along with strong continuum radiation appeared in the boron plasma spectrum, which did not permit observation of harmonics with wavelengths shorter than 65 nm.

High-order harmonic generation was optimized by varying the experiment geometry, with the femtosecond pulse focusing conditions varied from "weak" ($b \approx 4.8$ mm, $L_p \approx 0.6$ mm, $b > L_p$, where b and L_p are respectively the confocal parameter of the focused radiation and the plasma length) to "strong" ($b \approx 0.9$ mm, $L_p \approx 0.6$ mm, $b \cong L_p$). In this case, the order of generated harmonics somewhat increased (the 65th harmonic, $\lambda = 12.2$ nm), as did the conversion efficiency in the plateau region (5×10^{-7}).

An analysis of the harmonic radiation generated using the second (longitudinal) HHG scheme demonstrated the crucial role of the characteristics of heating-pulse radiation in the formation of optimal conditions for the frequency conversion of laser radiation. For a low-energy pulse, a power-law dependence with the exponent equal to 3.5 was observed for the dependence of high-harmonic intensities on the heating pulse intensity near the target. Up to 57th-order harmonics were generated in this investigation. However, the conversion efficiency remained low (10^{-7} [36]).

2.2.2. Silver

In the case of silver plasma, the plateau-like pattern was observed for harmonics of orders higher than nine (see Fig. 2.3). The conversion efficiency (2×10^{-6}) was higher than the conversion efficiency in the boron plasma under similar experimental conditions. The calibration of the recording equipment in the high- and low-order harmonic regions enabled absolute measurements of the conversion efficiency to be performed [29]. The subsequent optimization of this experiment led to an appreciable increase in the HHG efficiency in the plateau region (8×10^{-6}), which turned out to be comparable to the conversion efficiency in gaseous sources.

An important characteristic of this process is the dependence of the highest-generated harmonic order (harmonic cutoff, H_c) on the intensity of the femtosecond pulse (I_{fp}). For silver plasma, this dependence of $H_c(I_{fp})$ was linear and reached saturation for $I_{fp} = 3.5 \times 10^{14}$ W cm^{-2}. This intensity of

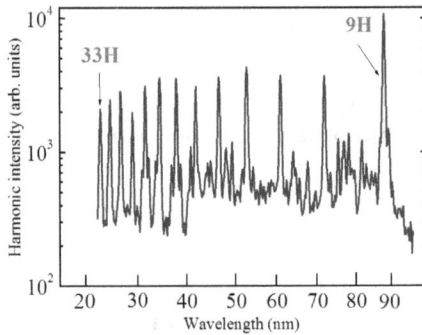

Fig. 2.3 High-order harmonic spectrum generated in silver plasma. Adapted from [29] with permission from Elsevier.

the 150 fs pulses used in experiments is much higher than the tunnel ionization intensity for neutral silver atoms. The results of studies show that the role of free electrons generated in the ionization of neutral atoms is insignificant, owing to their low density. Because the ionization of neutral atoms does not lead to a saturation of the above dependence and does not entail a substantial phase mismatch, self-defocusing, or suppression of the peak intensity in the plasma region, this saturation (the 57th harmonic) may be caused by the further ionization of singly charged ions and an increase in the free-electron density [25, 37].

Calculations of the tunnel ionization threshold for singly charged silver ions (2.1×10^{14} W cm^{-2}, $I_{2i} = 21.48$ eV) show that it is nearly the same as the intensity at which the dependence $H_c(I_{fp})$ is saturated. Subsequent experiments with silver plasmas, which were carried out using the 48 fs, 795 nm pulses, showed an increase in H_c (the 61st harmonic, $\lambda = 13$ nm, Fig. 2.4). The conditions for harmonic generation were optimal when the distance between the radiation–plasma interaction region and the target surface was in the range

Fig. 2.4 Distribution of harmonics generated in silver plasma using 48 fs pulses. Adapted from [37] with permission from Springer Science+Business Media.

of 100–200 μm, depending on the order of the harmonic generated. The optimal distance was dependent on the temporal delay between the pulses. These experiments, like the previous ones, exposed the adverse effect of an excessive heating picosecond pulse intensity (I_{pp}) on the target, resulting in a lowering of the conversion efficiency (for $I_{pp} > 3 \times 10^{10}$ W cm^{-2}) or a total loss of the HHG conditions (for $I_{pp} > 8 \times 10^{10}$ W cm^{-2}) [37]. An important goal of these investigations was to determine the optimal HHG conditions for different positions of the focal plane of the lens that focused the fundamental femtosecond beam into the silver plasma. The dependence of harmonic intensities on the location of the focus relative to the plasma plume was largely determined by the pulse energy and by the intensity at the focus of the lens in use. For a low intensity, a single maximum was observed in the dependence $I_h(z)$ (where I_h is the harmonic intensity and z is the longitudinal coordinate), occurring when the focal region coincided with the plasma plume. But for high femtosecond radiation intensity, this dependence had two peaks: the first, the smaller one, when the focal plane was located in front of the plasma plume, and the greater one when the focal plane was located behind the plume. Under this irradiation intensity, focusing into the plasma region gave rise to substantial additional ionization of the plasma plume and subsequent intense emission in the XUV range. In this case, the emergence of additional free electrons and phase self-modulation were responsible for the loss of phase matching between the fundamental radiation and harmonic waves and for a consequential sharp decrease in the HHG efficiency.

When the focal point is located behind the nonlinear medium, one would expect a change in the spatial shape of laser radiation (from a Gaussian distribution to a more flat distribution in the central region of the beam) due to higher free-electron density in the paraxial region, which leads to defocusing of this part of the beam. These conditions are analyzed in detail in [38]. When the nonlinear medium is located in front of the focal plane of the femtosecond pulse, the central part of the laser beam experiences negative nonlinear refraction caused by free electrons, which increases the divergence of the beam. At the same time, the "wings" of the spatial distribution of this beam are still focused into the plasma. In this case, it is possible to attain a phase-matched energy transfer from the pump wave to harmonic waves over a longer path in the plasma. This effect was earlier observed in gas media [39, 40].

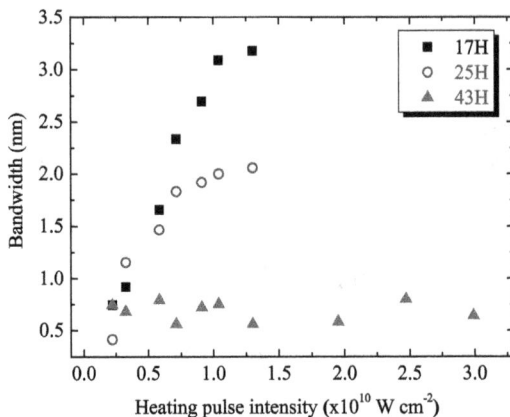

Fig. 2.5 Dependence of the spectral widths of the 17th, 25th, and 43rd harmonics on the intensity of heating picosecond pulses. Reprinted from [41] with permission from American Physical Society.

The importance of careful control over the heating pulse and driving probe pulse energies for efficient HHG was demonstrated in subsequent experiments with shorter (35 fs) pulses. Variations in spectral harmonic widths with increasing heating pulse intensity were observed in experiments with silver plasmas [41]. Figure 2.5 shows the spectral widths of the 17th, 25th, and 43rd harmonics as functions of the heating picosecond pulse intensity. The form of these dependences clearly shows that, while the spectral width of the lower-order harmonics increases with I_{pp}, this parameter remains invariable for higher-order harmonics. Plasma plume characteristics were calculated using the thermodynamic code HYADES. This one-dimensional code was devised for calculating properties of laboratory plasmas produced under the action of high-power energy sources. The code was elaborated to suit the following requirements: (i) convenience in use by an experimenter; (ii) recourse to simple but reliable approximations involving different physical models; (iii) the possibility of modifying it with relative ease and including new models for consideration; and (iv) the possibility of using different computer operating systems.

Recent modifications to this code have allowed more realistic values of plasma parameters to be calculated, in particular, of temperatures in the range of several electronvolts. Elaborated by Larsen and Lane [42], these

Table 2.1. Calculated characteristics of silver plasmas. Reprinted from [41] with permission from American Physical Society.

Heating pulse intensity $\times 10^{10}$ W cm^{-2}	0.3	0.59	0.9	1.3	1.95	2.6	3.25
Free electron density $\times 10^{17}$ cm^{-3}	0.89	2.56	3.2	4.0	5.79	6.33	8.0
Ion density $\times 10^{17}$ cm^{-3}	0.89	2.56	3.17	3.41	4.45	4.55	5.06
Ionization degree	0.5	0.7	1.01	1.17	1.3	1.39	1.58

modifications of the code HYADES were later used to calculate the dynamics of deposition using a plasma plume [43], to model different plasma instabilities [44], to optimize the characteristics of tenuous plasmas for various applications [45], etc. Experiments in harmonic generation from silver plasmas and harmonic brightness variations (in other words, spectral width variations), as well as calculations by the code HYADES show that initially (for a low heating pulse intensity, $I_{pp} < 0.9 \times 10^{10}$ W cm^{-2}), the degree of plasma ionization lies in the range from 0.5 to 1 (Table 2.1 [41]). In this case, the plasma consists of neutral atoms and singly charged ions responsible for HHG. For pulse intensities in the range $(0.9–3) \times 10^{10}$ W cm^{-2}, the ionization level becomes higher than 1. Under these conditions, the plasma consists primarily of singly charged silver ions and a small fraction of doubly charged ions. In this case, spectral broadening of lower-order harmonics is observed, while the spectrum of higher-order harmonics remains invariable. For intensities above 3×10^{10} W cm^{-2}, the ionization level exceeds 1.5. In that case, an increase in the free-electron density leads to decrease of HHG conversion efficiency due to a phase mismatch of the participating waves.

The technique for the calibration of harmonic conversion efficiency was as follows [29]. The low-order harmonics were analyzed using the "monochromator (Acton, VM502) + sodium salicylate + photomultiplier tube" detection system. At the first step, the third-harmonic signal was measured by this system using both the known energy of the third harmonic of 796 nm radiation generated in nonlinear crystals and the third harmonic generated in the silver plasma. This calibration gave the absolute conversion efficiency for the third harmonic generated from plasma as 5×10^{-4}. The monochromator allowed observation of the low-order harmonics up to the spectral range of 70 nm. Each of the next low-order harmonics was gradually decreased with a coefficient of 4 to 6, and the conversion efficiency for the ninth harmonic (88.4 nm) was measured to be 3×10^{-6}. Then the same ninth harmonic was observed using the "extreme ultraviolet spectrometer + MCP + phosphor + CCD" detection system. Starting from the 11th harmonic, the harmonic spectra showed a plateau pattern. The harmonic output in the plateau region was improved using the optimal heating pulse energy, delay between pulses, and target–beam distance, and achieved the conversion efficiency of 8×10^{-6}.

2.2.3. Gold

Analysis of laser-produced plasma characteristics plays an important role in the optimization of frequency conversion of high-power laser radiation. One such characteristic is the spectral composition of plasma in the visible and UV ranges. Early in the investigation of HHG in laser-produced plasma, the spectrum was analyzed from time-integrated spectral characteristics of the plasma plume. To state this in different terms, the recorded spectrum of the plasma was accumulated during the whole period of its irradiation [46]. However, recent studies of HHG in the plasma produced on the surface of several targets involved a technique of analysis of the plasma plume at different stages of its development [31]. This revealed several features in the dynamics of plasma over-excitation under irradiation of different targets. In what follows, we show the results of an analysis of the plasma produced on the surface of gold, which was performed for the purpose of optimizing its characteristics for HHG. More details of this technique are discussed in Section 6.2. The interval between each spectrum measurement was 20 ns, beginning from the onset of gold plasma excitation to the moment at which the plasma appreciably recombined under the conditions optimal for HHG (150 ns). These investigations were carried out for both "optimal" and "nonoptimal" plasmas. The term "optimal" characterizes the plasma plume state that corresponds to the maximum HHG efficiency. In a nonoptimal plasma, the conversion efficiency is significantly lower due to the effects of several limiting factors. The plasma spectrum was measured in a narrow UV range (266–295 nm), in which it was possible to trace the excitation of singly charged ions. The optimal plasma afforded the generation of a plateau-like harmonic distribution, the generation of highest-order harmonics being observed in this case.

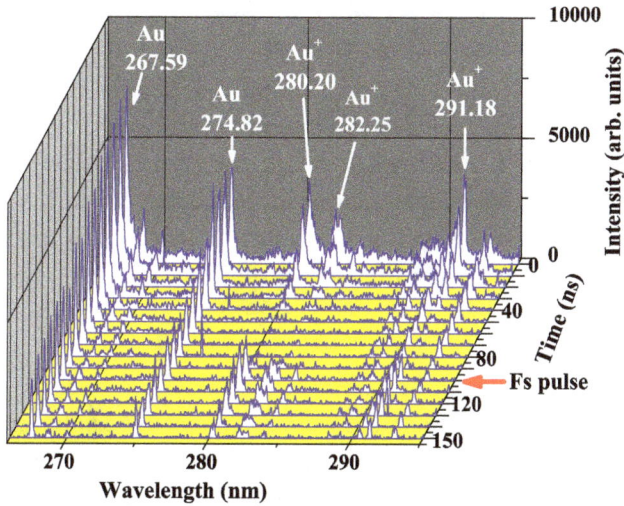

Fig. 2.6 Dynamics of the UV emission spectra of gold plasma, $I_{pp} = 1 \times 10^{10}$ W cm^{-2}. Reprinted from [28] with permission from American Institute of Physics.

Figure 2.6 shows the dynamics of the UV spectrum of gold plasma irradiated by 35 fs pulses 100 ns after the onset of plasma development. The decay times for the lines of excited neutral atoms were appreciably different from the decay times of ionic lines. It can be seen from Fig. 2.6 that the intensities of Au II lines (280.20 nm, 282.25 nm, and 291.18 nm), which are excited under these conditions, decrease rapidly, while the emission in the lines arising from neutral gold atoms (267.59 nm and 274.82 nm) lasts sufficiently longer than the emission of ionic lines. It can also be seen from these spectra that the femtosecond pulse, which arrived 100 ns after the onset of plasma formation, primarily excites the ionic lines, while the lines of neutral atoms do not undergo significant changes [28].

A different picture was observed in the UV spectrum dynamics, with an increase in the heating pulse intensity at the target surface. Increasing this parameter from 1×10^{10} W cm^{-2} to $(2–3) \times 10^{10}$ W cm^{-2} resulted in a substantial strengthening of the plasma emission. Along with an approximately four-fold increase in ionic line intensities, there emerged other spectral features arising from additional ion transitions in the spectrum of gold. The UV plasma spectra for optimal and nonoptimal harmonic generation conditions, 90 and 100 ns after the onset of plasma formation, were significantly different.

A similar picture was also expected for the spectral distribution in the XUV spectral range. The experimental conditions did not allow analysis of the dynamics of the plasma spectrum with temporal resolution in the XUV range, but the emergence of intense ionic lines in the domain of the plateau-like harmonic distribution, which was seen from the integral spectra of harmonics and plasma, led to a substantial decrease in the intensity of the harmonics and sometimes to their complete disappearance. In particular, a relatively small increase in the heating pulse intensity at the target surface from 1×10^{10} $W\,cm^{-2}$ to $2.5 \times 10^{10}\,W\,cm^{-2}$ causes a 2.5-fold decrease in the intensity of the 13th harmonic.

This analysis allowed optimization of HHG by selecting the most appropriate intensities for the heating pulse and the probe pulse, as well as the delay between them. This optimization enabled reduction of the effect of the main HHG limiting factor in radiation conversion in plasmas — the high free-electron density, which is responsible for phase mismatch and self-defocusing. Optimization of this kind permitted the harmonics generated in the gold plasma to advance towards shorter wavelengths. Under these conditions, the generated harmonic order was as high as 53 ($\lambda = 15.09$ nm). A comparison of the conversion efficiencies for silver and gold plasmas showed that in the latter case the intensities of the 25th to 49th harmonics were several times lower. The conversion efficiency in the plateau region in gold plasma was equal to 2×10^{-6}.

2.3. Application of 400-nm Radiation for Harmonic Generation in Laser Plasma

The application of shorter-wavelength sources of driving radiation for HHG has some attractions. The important feature of using shorter-wavelength photons for HHG is the smaller mismatch between the driving radiation and harmonic waves due to the smaller influence of the free electrons on the phase matching conditions. For the high-order harmonics, the phase mismatch in plasma is determined mostly by free electrons ($\Delta k = (n^2 - 1)(\omega_p)^2 / 2nc\omega$, where ω_p is the plasma frequency, n is the order of the harmonics, c is the speed of light, and ω is the laser frequency). At a free-electron density of $2 \times 10^{17}\,cm^{-3}$ (i.e., at an ionization rate of 1 for some ablation conditions of metal targets), the phase mismatch is of the order of $90\,cm^{-1}$ (for 400 nm driving pulses). Thus, the coherence length for harmonics up to the 21st order (~19 nm) is approximately 0.7 mm, which is comparable to or higher than the plasma dimensions (0.3–0.6 mm). At the same time, for 800 nm radiation, the same free-electron density creates an acceptable phase mismatch only for harmonics above 38 nm. These estimates show that the use of shorter-wavelength sources is advantageous to satisfy the phase matching conditions for shorter wavelengths of harmonic radiation in the case of the influence of free electrons (at the same concentration of electrons). At the same time, the three-step model describing HHG in gaseous and plasma media predicts that the cutoff energy E_c would drop with the decrease of the wavelength of fundamental radiation by a factor of λ^2. Therefore, the cutoff harmonic order should scale as λ^3. It is important in analyzing the application of shorter-wavelength sources compared to longer-wavelength ones to define the advantages and disadvantages of conversion efficiency through wavelength control.

The first observations of HHG from a laser surface plasma excited by subpicosecond short-wavelength radiation (KrF excimer laser, $\lambda = 248$ nm) were reported in [16, 17]. The authors observed harmonics up to the 21st order ($\lambda = 11.8$ nm) generated in a lead plasma [17]. They also analyzed the positive phase mismatch caused by free electrons in the cases of IR and UV driving radiation and found a considerable difference in these two cases, showing a significantly lower phase mismatch in the XUV range when using the UV pulses. The same conclusions were reported in early studies of HHG in gas jets [47].

Some time ago, a two-color (fundamental and second harmonic) scheme was proposed and realized for gas HHG to increase the harmonic yield and to generate single attosecond pulses. For these purposes one has to add a weak second-order field to the fundamental beam [48]. Recently, various schemes using a second-harmonic pump for phase matching and improvement of electron trajectories have been proposed [49, 50]. Similar studies could be performed using plasma HHG to achieve the attosecond timescale. For these purposes one has to learn more about the properties of the harmonics generated from the 400 nm pump laser itself.

In this section, the application of 400 nm radiation for HHG from various plasma plumes created on the surfaces of solid-state targets is discussed. The experimental scheme was described in detail in the previous chapter. To create the plasma plume, a heating pulse from the chirped radiation of a Ti:sapphire laser ($t = 210$ ps, $\lambda_{pp} = 800$ nm) was focused on a target placed in the vacuum chamber. After some delay (40 to 100 ns), the femtosecond main pulse ($E = 5$ mJ, $t = 35$ fs, $\lambda_{fp} = 400$ nm central wavelength, 9 nm FWHM bandwidth) was focused on the plasma from the orthogonal direction. This radiation was produced during second-harmonic generation of the 800 nm radiation of the Ti:sapphire laser in the nonlinear crystals. The generated harmonics were analyzed using an XUV spectrometer.

The measurements of harmonic cutoff (H_c) for the 400 nm laser pumped silver plasma showed a four-fold decrease compared with the case of the 800 nm pump, instead of the expected eight-fold decrease ($H_c \propto \lambda^3$). Comparison with previous studies of harmonic efficiency in the plateau range in the case of HHG of the 800 nm radiation in silver plasma (8×10^{-6}) showed a ten-fold decrease of conversion efficiency in the case of the 400 nm radiation, probably due to lower intensity of UV pulses. For the 800 nm pump, the cutoff order from the silver plasma was measured to be in the range of the sixties harmonics [29, 37, 41]. An analogous tendency was observed for an aluminum plasma [27] (Fig. 2.7), when the harmonic cutoff (15th harmonic) considerably exceeded the predictions of the three-step model. The H_c in the aluminum plasma obtained from the 800 nm laser (43rd harmonic) was only three times higher than in the case of doubled-frequency pump radiation. The discrepancy between the theory and experiment was also observed in the

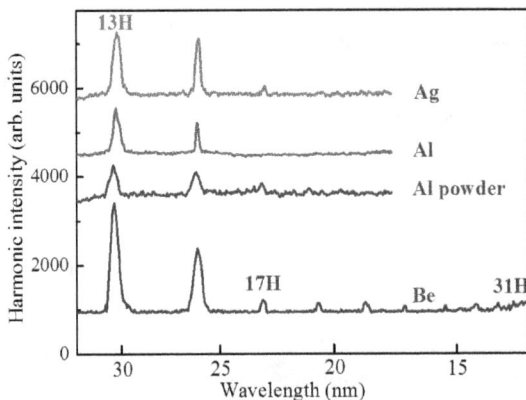

Fig. 2.7 Harmonic spectra obtained in silver, aluminum, aluminum powder, and beryllium plasmas in the case of 400 nm driving radiation. Reprinted from [27] with permission from American Institute of Physics.

case of the ablation of aluminum powder, when the 17th and higher orders appeared in the harmonic spectrum.

The reason for such a discrepancy between the three-step model and the experiments using shorter-wavelength radiation can be attributed to the different species (atoms, singly and doubly charged ions) that contribute to the harmonic generation of the 800 nm and 400 nm pump lasers. One should also keep in mind that, since the harmonic cutoff energy scales linearly with laser intensity, it is important that the pump laser intensity in the medium was kept the same for the two pump lasers when one compares these two cases. Another reason is that this scaling neglects the ionization potentials of particular species, since the scaling is only valid for high-order harmonics when the ionization potential becomes less important. Also, the difference in the ionization-induced defocusing in the cases of the 800 nm and 400 nm pulses could play a role in the unexpected scaling of the harmonics. Less influence of free electrons on the phase matching conditions for the 400 nm pump laser could also be a reason for departing from the λ^3 scaling. The plateau pattern was not so pronounced in the case of the 400 nm driving radiation compared to the 800 nm laser pulses, when the harmonic distribution from most plumes showed almost equal intensities for the harmonics exceeding the 13th order. The only sample where the plateau pattern appeared in the case of the 400 nm driving radiation was the beryllium plasma. The harmonics between the 17th and 31st orders showed a clear plateau-like pattern (Fig. 2.7). Some plasma samples (tin, manganese, chromium, antimony) demonstrated resonance-induced enhancement of single harmonics. Such enhancement occurred when the harmonic wavelength was in the vicinity of a radiative transition possessing strong oscillator strength. This process will be discussed in detail in the following chapter. Here we just mention a few studies where 400 nm probe pulses were used for the observation of resonant enhancement.

Studies using 400 nm radiation confirmed previously reported peculiarities of the harmonic spectra from a chromium plasma in the vicinity of the short-wavelength wing of the strong spectral band of the 3p $\rightarrow$ 3d transitions of Cr II ions [27, 51] (Fig. 2.8). The observed three-fold enhancement of the 15th harmonic of 400 nm driving radiation can be attributed to the enhancement of the nonlinear susceptibility of this harmonic induced by the influence of the same transitions, though not so pronounced compared with the 29th harmonic

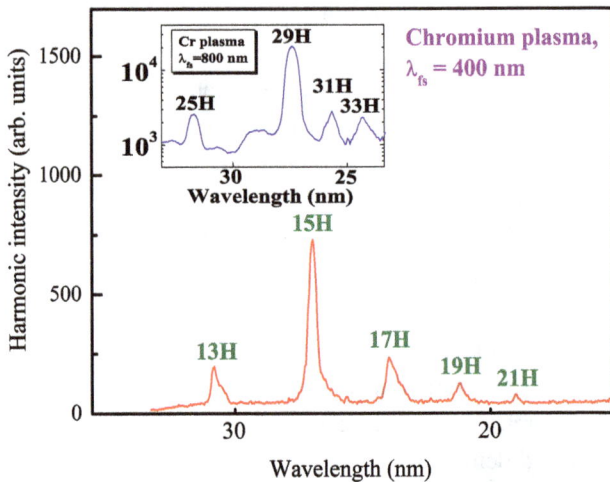

Fig. 2.8 Single-harmonic enhancement obtained in the case of 400 nm radiation propagating through chromium plasma. Inset shows the enhancement of single harmonic obtained in the case of the 800 nm probe radiation. Reprinted from [27] with permission from American Institute of Physics.

of 800 nm radiation (see inset in Fig. 2.8). In the case of manganese plasma, the maximum harmonic order (21st harmonic) was considerably lower compared with the case of the 800 nm pump (101st harmonic). A six-fold enhancement of the 17th harmonic was observed in these studies (Fig. 2.9). The single-harmonic enhancement in the case of the 400 nm pump was in stark contrast with the multiple-harmonic enhancement observed in the case of the 800 nm driving pulses (see inset in Fig. 2.9). In the latter case, the harmonics between the 33rd and 41st orders demonstrated three-fold enhancement, though not so strongly as in the former case (six-fold). The same features were observed in the case of tin plasma (Fig. 2.10).

If an accelerated electron recombines with the core, it can fall into either the ground state or the excited state from which it originates. The probability of recombining into these states is determined by their oscillator strength. That is why, in the case of excited states with weak oscillator strength, *gf*, a weak multiple-harmonic enhancement attributed to recombination into the ground state could be observed, while in the case of strong *gf* of a single excited state, one can notice prevailing recombination into this state with a subsequent transition into the ground state emitting a single harmonic.

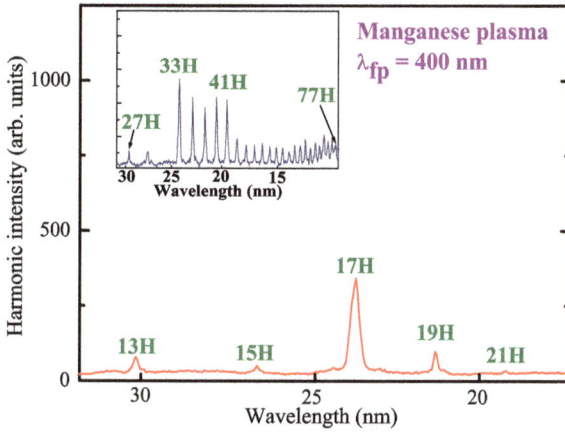

Fig. 2.9 Single-harmonic enhancement obtained in the case of 400 nm radiation propagating through a manganese plasma. Inset shows the enhancement of multiple harmonics obtained in the case of 800 nm probe radiation. Adapted from [27] with permission from American Institute of Physics.

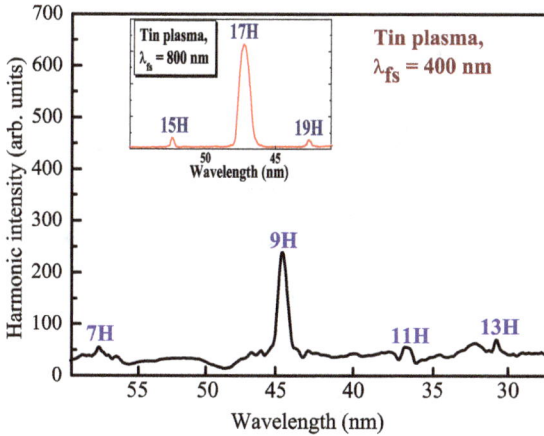

Fig. 2.10 Single-harmonic enhancement obtained in the case of 400 nm radiation propagating through a tin plasma. Inset shows the enhancement of the single harmonic obtained in the case of 800 nm probe radiation. Adapted from [27] with permission from American Institute of Physics.

The HHG studies at different pump laser chirps were performed to tune harmonic wavelengths in proximity to ionic transitions possessing strong oscillator strengths. The chirp of an 800 nm laser pulse was varied by adjusting the distance between the two gratings of the pulse compressor. Varying the laser chirp resulted in a considerable change in the harmonic distribution from laser plasma in the case of HHG using 800 nm chirped pulses. However, in the case of 400 nm radiation, the control of chirp conditions appeared to be not so efficient for manipulating the single-harmonic conversion efficiency. The enhancement of single harmonics of 400 nm radiation in the case of the chirp variations of 800 nm radiation did not change with variation in the pulse duration and sign of the chirp.

Figure 2.11 presents a summary of the studies of the enhancement factor at different chirps and pulse durations of the driving radiation. In the case of manganese and tin plasmas, the enhancement factors (i.e., the ratios between the enhanced harmonic and neighboring harmonic intensities (I_{EH}/I_{NH})) remained approximately the same in a broad range of variations of chirp and pulse duration, contrary to the variations of the enhancement factors

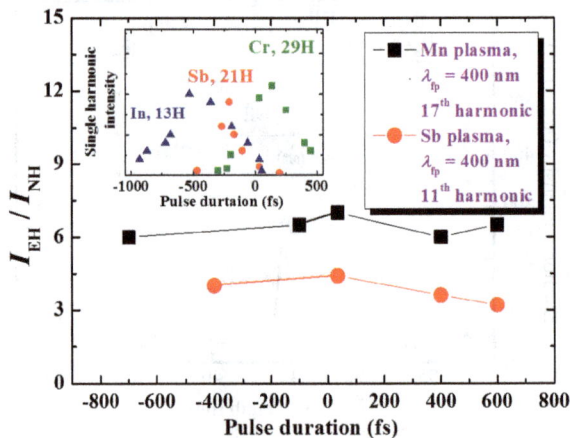

Fig. 2.11   Ratio of the enhanced harmonic and neighboring harmonic intensities obtained in manganese and antimony plasmas at different chirps and pulse durations of the 400 nm driving radiation. Inset: Single-harmonic variations at different chirps of the 800 nm radiation in the cases of indium (triangles), chromium (squares), and antimony (circles) plasmas. Positive and negative signs of the pulse durations correspond to positive and negative chirps of the driving radiation. Reprinted from [27] with permission from American Institute of Physics.

for different chirps and pulse durations in the case of the 800 nm pump (see the inset in Fig. 2.11). The intensity of the 17th harmonic of 400 nm radiation generated in manganese plasma remained strong in a broad range of chirp and pulse duration variations and it was difficult to detune this harmonic from the quasi-resonance conditions [27]. Such a feature can be explained by the relatively narrow bandwidth of the 400 nm radiation (~9 nm), which allowed the tuning of the 17th harmonic within a narrow spectral range (0.3 nm) that was insufficient to detune the harmonic wavelength far from the resonance transition responsible for the enhancement of this harmonic. The same can be said about the experiments with antimony plasma (Fig. 2.11). Note that, in the case of 800 nm radiation, the variation of laser chirp allowed for a considerable change in the enhancement of specific harmonics [27, 51–53] due to the broad bandwidth of the pump radiation (40 nm). In the meantime, the tuning of harmonic wavelength in the case of the 400 nm pump could be achieved by tuning the phase matching conditions in the KDP crystal used for second-harmonic generation of the 800 nm radiation. As mentioned above, the coherence length ($L_c = \pi/\Delta k$) for the shorter-wavelength pump is expected to be a few times longer for the same spectral range, which can be considered as an advantage of using the 400 nm pump instead of 800 nm radiation. This allows an increase in the plasma length (L_p) in the former case, thus enhancing the harmonic yield, which is proportional to $(L_p)^2$. Some other advantages of the shorter-wavelength source include higher harmonic-conversion efficiency at specific spectral ranges. In particular, the low-order harmonics from the shorter-wavelength source correspond to the mid-order harmonics of the 800 nm pulses. This means that, in the range of 100–200 nm, the harmonic energy from the 400 nm driving pulses will be higher than that from the 800 nm pump. Shorter pulse duration and better coherence properties of the second harmonic of 800 nm radiation could also be advantageous for HHG. This is because the improved phase conditions of the pump radiation result in improved phase matching conditions between the harmonic and the driving waves. This can lead to further shortening of the pulse duration for harmonics generated by the 400 nm pump. However, the relatively small second-harmonic conversion efficiency in the thin nonlinear crystals ($\leq 20\%$) still remains an obstacle for application of the 400 nm pump in HHG studies.

2.4. High-Order Harmonic Generation in Plasmas Produced by Laser Pulses of Different Durations

The results of the investigations outlined in previous sections of this chapter demonstrate the decisive role of plasma characteristics in attaining the highest efficiency of coherent radiation conversion in the XUV range as well as in generating the highest-order harmonics. For this, different techniques have been used, such as (i) an analysis of the ionization state and of electron and ion densities using the code HYADES [41]; (ii) investigations of integral laser-produced plasma spectra [37, 46] and of time-resolved plasma spectra with the use of a high-speed technique allowing the laser plasma to be analyzed with a high temporal resolution [21, 28]; (iii) measurements of the divergence of a femtosecond beam transmitted through the laser plasma [54]; and (iv) an analysis of nonlinear optical plasma parameters using the Z-scan technique [54].

Several important characteristics, such as the duration of the heating pulse, were not appreciably varied during plasma production. The overwhelming majority of investigations of plasma HHG were carried out using pulses several hundred picoseconds long. The conclusion drawn in such cases involved determination of the "optimal" heating pulse intensity at the target surface ($\sim(0.5-3) \times 10^{10}$ W cm^{-2}, depending on the physical material properties [41, 46, 52], that afforded the highest value of H_c as well as the highest attainable HHG efficiency in the plateau domain. In this connection, it seems expedient to undertake a comparative analysis of HHG in plasmas produced by pulses of various durations: these investigations could allow determination of which parameter of the heating radiation (the pulse energy, fluence, or the radiation intensity at the target surface) plays a crucial role in optimal plasma formation.

Another interesting point is the effect of the target material atomic number Z on the HHG efficiency for different delays between the participating pulses. It is important to determine whether the target characteristics affect the optimal delay between the heating pulse and the femtosecond pulse. It may be expected from general considerations that the HHG dynamics would be different for "light" and "heavy" target atoms (i.e., for low- and high-Z materials), because the heating, melting, evaporation, and escape from the target surface, as well as the subsequent cooling and recombination of target particles depend on the atomic weight and target density. The difference in the escape velocity and recombination between light and heavy atoms and ions of laser-produced plasmas may have a pronounced effect on the plasma density characteristics at different distances from the target. In this case, optimizing the delay between the heating and probe pulses would favor the attainment of better characteristics of radiation conversion for light or heavy targets.

In this section, we analyze the results of investigations of HHG in the plasmas produced on the surface of low-, medium-, and high-Z targets irradiated by pulses varied over five orders of magnitude in duration (from 160 fs to 20 ns). These investigations show that plasma formation is crucial in the optimization of HHG, and the optimal plasma characteristics are largely determined by the pulse energy, while the radiation intensity at the target surface affects the dynamics of harmonic generation to a much smaller degree. In the course of these experiments, it was possible to reveal the differences in HHG in plasmas produced at the surfaces of targets consisting of light or heavy elements, when the delay between the pulses becomes the controlling factor for the attainment of efficient harmonic generation [55].

The setup of the experiments described below was similar to the previous ones. Initially, a 210 ps heating pulse was used. The duration of this pulse was varied by compression in an additional compressor. In particular, in the HHG research in this case, the compressed 160 fs and 1.5 ps pulses were used as the heating pulses as well. The energy of these pulses was maintained at about the same level (10 mJ). The pulse was focused onto the target using a spherical (or cylindrical) lens. Materials with high (silver, barium), medium (zinc, nickel, manganese), and low (graphite, boron, lithium) atomic numbers were used as the targets.

The intensity of 210 ps pulses at the target surface was kept in the interval $I_{pp} \approx (1-5) \times 10^{10}$ W cm^{-2}. In the cases of 160 fs and 1.5 ps pulses, the radiation intensity at the target surface was much higher, because these investigations were carried out under the same experimental geometry and the energies of pulses of different durations were about the same.

After some delay, the focused radiation of the probe pulse (115 fs, 20 mJ, 795 nm) passed through the produced plasma. Because the highest attainable intensity (6×10^{16} W cm^{-2}) of this pulse in the focal plane was much higher than the tunnel ionization thresholds for singly and doubly charged plasma ions, the position of the focal region (behind or in front of the plasma) was chosen in such a way as to maximize the harmonic intensities. For this position, the intensity of the probe pulse in the plasma plume was in the range $7 \times 10^{14} - 1 \times 10^{15}$ W cm^{-2}. These experiments were performed under "weak" focusing conditions ($b > L_p$, where b is the confocal parameter and L_p is the plasma length).

Figure 2.12(a), (b), (c), shows the results of investigations of H_c as a function of the pulse duration for heavy (silver ($Z = 47$)), intermediate (zinc ($Z = 30$)), and light (boron ($Z = 5$) and lithium ($Z = 3$)) targets. These experiments showed that there is a substantial difference in harmonic characteristics among these three groups of targets. For heavy targets (and for the majority of medium-Z targets), no variation in the highest generated harmonic order was observed when 160 fs, 1.5 ps, and 210 ps pulses were used for plasma formation. In several cases, the 20 ns pulses of the second harmonic of an yttrium–aluminum garnet laser were used as a heating pulse as well. The use of these pulses did not entail substantial changes (in the case of heavy targets) in H_c.

However, an essential dependence of H_c on the duration of the heating pulse was observed in the case of light targets. But even in these circumstances, no fundamental limiting factors were revealed that would lead to a strong violation of HHG conditions for a substantial variation of the heating radiation intensity, which amounted to five orders of magnitude. The only exception was the lithium plasma, in which high-order harmonics were generated only with the use of long (210 ps) heating pulses and short delays. In this case, the main distinction in HHG characteristics consisted in the delay between the heating and probe pulses. These investigations demonstrated that the fluence of heating pulse energy plays the decisive role in the formation of optimal conditions for plasma HHG, rather than the intensity of this pulse on the target surface.

The integral spectra of laser-produced plasmas in the visible and UV spectral ranges showed that the general spectral line distribution was about the same despite the difference in duration of the pulses that generate the plasma on the target surfaces. The manganese and silver plasma emission spectra in the visible and UV ranges, excited by 160 fs, 1.5 ps, and 210 ps pulses, showed that the duration of the heating pulse had no appreciable effect on the dynamics

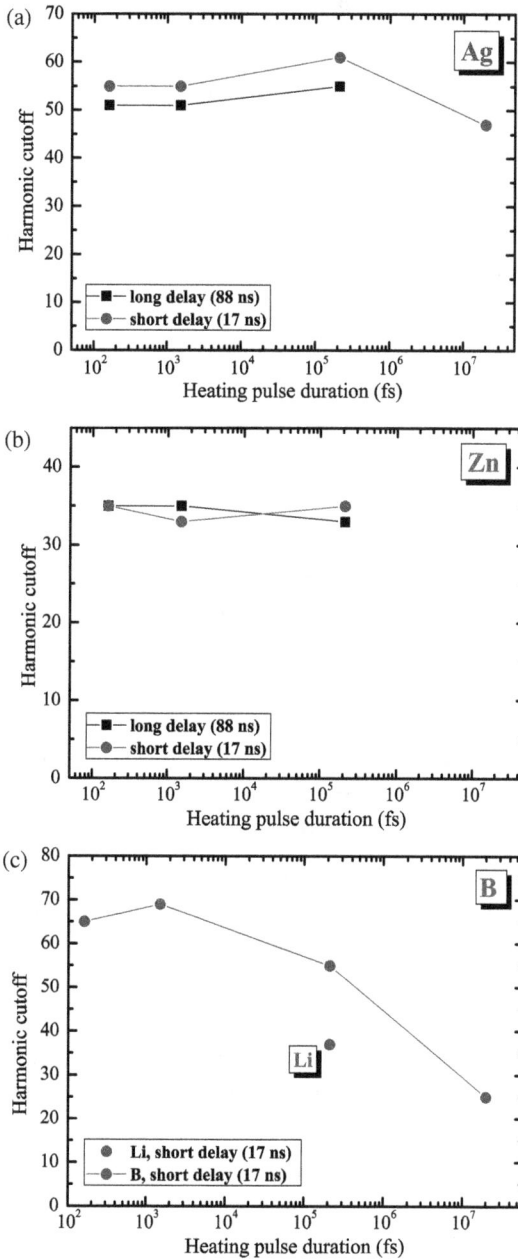

Fig. 2.12 Highest harmonic orders generated in (a) heavy (silver), (b) intermediate (zinc), and (c) light (lithium and boron) target plasmas for different delays and heating pulse durations. Adapted from [55] with permission from American Physical Society.

of laser plasma emission and that the leading role in plasma excitation was played by the pulse energy. The same is true for plasma emission in the XUV range, which is an indication that the laser plume characteristics are about the same for substantially different pulse intensities at the target surface excited by radiation of different pulse durations, once we achieve the conditions of "optimal" plasma formation for efficient HHG.

The HHG efficiency depended on the delay between the heating and probe pulses. For very short delays (5 ns and less), harmonic generation was not observed due to the insufficient density of the nonlinear medium in the region of interaction with the femtosecond pulse. At longer delays, a substantial difference between the harmonic generation properties of light and heavy targets appeared. For a 17 ns delay, HHG was observed for all the targets under investigation, while for an 88 ns delay, harmonics were generated only in plasmas from the high- and medium-Z targets (see Fig. 2.12). For this long delay, harmonics were not generated in the plasmas of light targets (with the exception of graphite). We note that the values of H_c for high- and low-Z targets were about the same for short and long delays between the pulses.

This difference in generation properties may be related to the dynamic plasma characteristics for light and heavy targets. Light particles fly away from the target at a higher velocity, which may entail a lower laser-produced plasma density in the region of interaction with the probe pulse several tens of nanoseconds after the onset of plasma formation. But heavier particles stay near the target (where the laser radiation frequency is converted) for a longer time. For heavy targets, the delay between the pulses turned out to have a weaker effect on the generated harmonic intensities than for light targets. Therefore, in the search for possible targets, their physical parameters should be taken into account.

It was noted above that the second ionization potential of targets plays an important role in determining the highest generated harmonic order. The results of investigations described in this section revealed additional target and pulse parameters, as well as the delays between the pulses, which should be taken into account in the subsequent identification of materials whose plasmas favor the generation of highest-order harmonics with the highest conversion efficiency. From the empirical formula that relates the highest generated harmonic order to the ionization potential (I_i) of the particles participating in the conversion of laser radiation frequency ($H_c \approx 4I_i - 32.1$ [56]), it follows that the highest H_c is to be observed in a medium having the highest ionization potential of these particles. In the majority of cases, when singly charged ions played the role of the main component participating in HHG, the highest H_c values were observed in the plasma plumes of the materials with the largest value of the second ionization potential (in particular, in silver and boron plasmas). At the same time, H_c was seen to exceed the value expected from the above relation, which may be attributed to the participation of doubly charged ions in the frequency conversion. The highest harmonic orders observed in this case and the third ionization potentials of these materials (vanadium and manganese) fit the above relation satisfactorily. Hence, it follows that the plasma media that favor the realization of conditions for the generation of ultimately high-order harmonics should be sought among the materials with the highest third ionization potential.

The targets used in the investigations described in this section were chosen based on precisely these considerations. But attempts to attain H_c values exceeding the highest harmonic orders (the 71st harmonic for vanadium and the 101st harmonic for manganese plasmas) reported previously were unsuccessful. The reason may lie with specific phase matching relations in different plasma media, which may lead to efficient HHG with the participation of doubly charged ions only in a limited number of materials, while in other plasma plumes with a high value of the third ionization potential, the effect of an excess of free-electron density may largely impair the phase matching in the laser radiation frequency conversion. Laser-produced plasma is a highly complicated medium as regards its nonlinear optical properties. The behavior of this plasma system is appreciably subject to changes involving variations in the ionization state, ion and electron densities, ionization potentials, and a number of thermodynamic parameters. The processes defining HHG efficiency in this medium are rather complex and involve several additional factors that are nonexistent in the case of HHG in gases. In particular, the plasma medium initially consists of neutral particles, ions, and free electrons, and the densities of these plasma constituents depend to a great extent on the heating pulse and target characteristics. As noted above, an excess of free-electron density (generated, in particular, in the subsequent ionization of singly charged ions) noticeably impairs the phase matching between the harmonics and the radiation under conversion.

2.5. Analysis of Laser-Produced Plasma Characteristics for Optimization of HHG

An ion medium modifies the temporal characteristics of propagating femtosecond pulses owing to phase self-modulation. Furthermore, the spectral structure of high-order harmonics depends critically on the frequency modulation (chirp) of the radiation under conversion. Available from the literature are reports of investigations of the dynamics of chirped pulse propagation through ionized gas-jet sources and data about the spatial, spectral, and temporal parameters of laser radiation transmitted through such a medium, as well as the parameters of harmonics (see [57] and the references therein). Among the processes that affect the parameters of laser radiation, a prominent role is played by self-defocusing. Because self-defocusing appears to be one of the main limiting mechanisms for HHG (both in gases and in laser-produced plasmas), it was vital to analyze this process in greater detail under conditions close to the "optimal" plasma, in which HHG is realized most efficiently, as well as to determine the experimental conditions under which the self-defocusing effect is less significant [54].

In the experiments discussed below, the spatial distribution of the probe radiation ($t = 50$ fs) transmitted through the plasma plumes of different densities (from 5×10^{16} cm^{-3} to 2×10^{17} cm^{-3}) and laser radiation intensities was analyzed using a CCD camera in the far-field zone. The nonlinear optical characteristics of indium and molybdenum plasmas were investigated by the Z-scan technique. The spatial distribution of the radiation transmitted through the plasma remained invariable for low intensities of the femtosecond pulses even when the ionization threshold of target atoms was exceeded. This observation showed that the density of free carriers generated in the course of plasma formation and during the subsequent ionization of neutral plasma particles induced by a femtosecond pulse is insufficient for the self-defocusing of the laser beam. A different spatial distribution pattern of the laser radiation transmitted through the plasma was observed when the thresholds for tunnel ionization of singly and doubly charged ions were exceeded. The respective tunnel ionization thresholds for Mo I, Mo II, and Mo III are equal to 1.2 $\times 10^{13}$, 7×10^{13}, and 2.4×10^{14} W cm^{-2}. For $I_{fp} \approx 5 \times 10^{14}$ W cm^{-2}, the density of free electrons that emerged in the focal region in a time much shorter than the duration of the femtosecond pulse proved to be sufficient for the occurrence of self-defocusing of the laser beam. An annular structure was observed in the profile of the transmitted beam in this case, which was indicative of variations in the refractive index of the medium along the beam propagation axis (Fig. 2.13).

The Z-scan of different plasma plumes was performed similarly (Fig. 2.14). For low intensities of laser pulses, the normalized transmittance of the laser radiation transmitted through the plasma and the limiting far-field aperture

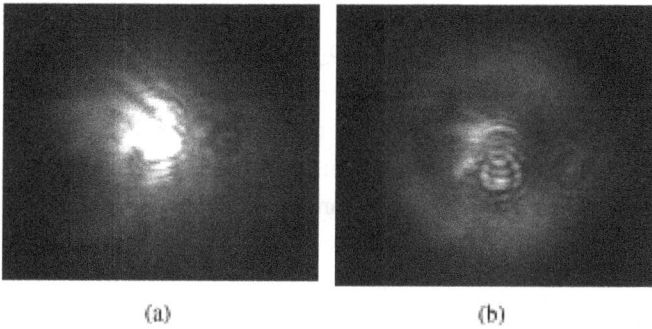

(a)   (b)

Fig. 2.13   Spatial distributions of femtosecond radiation after the propagation of a molybdenum plume at different intensities at the focal plane. (a) 4×10^{13} W cm^{-2}, (b) 5×10^{14} W cm^{-2}. Reprinted from [54] with permission from Optical Society of America.

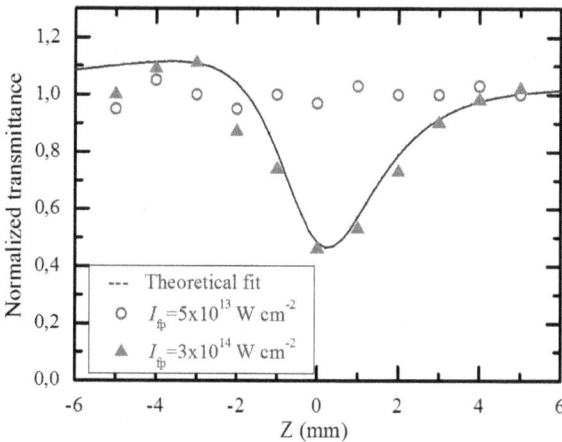

Fig. 2.14   Z-scan of indium plasma for low (circles) and high (triangles) intensities of femtosecond pulses. The solid curve represents the data of theoretical calculations. Reprinted from [54] with permission from Optical Society of America.

remained invariable. However, with an increase in the pulse intensity, a specific feature showed up in the normalized transmittance, which was a small peak with a subsequent substantial lowering of transmittance through the aperture. This feature was due to the self-defocusing and nonlinear absorption effects arising from the tunnel ionization of neutral particles and ions. The setup with a limiting aperture enabled determination of the magnitude and sign of the nonlinear refractive index γ as well as the magnitude of the nonlinear absorption coefficient β of the plasma plumes. In particular, for indium plasma, the γ and β were respectively equal to -2×10^{-18} cm^2 W^{-1} and 5×10^{-13} cm W^{-1} [54]. The β values depended to a large extent on the radiation intensity in the plasma plume region, which was to be expected considering the multiphoton nature of the nonlinear absorption. In particular, four-photon and five-photon absorption were the dominant processes in indium ($I_i = 5.8$ eV, $E_{ph} = 1.56$ eV) and molybdenum ($I_i = 7.1$ eV) plasmas. The four-photon absorption coefficient calculated for indium plasma was 5×10^{-42} cm^5 W^{-3}.

Investigations of the shape of the laser beam transmitted through the plasma at different radiation intensities and plasma plume densities allowed determination of the role played by the free electrons, which are generated during the ionization of neutral particles and ions, in changing the spatial and phase characteristics of the laser pulse in the focal plane. Self-defocusing and the corresponding changes of the phase characteristics of laser pulses do not permit preservation of the phase matching between the harmonic and probe radiation waves throughout the path of the pulse in a nonlinear medium. The intensity of laser radiation also decreases in this case, which is caused by an increase in divergence on the beam axis in the interaction region. These effects prevent attainment of favorable conditions for efficient HHG in the passage of laser radiation through the plasma, which lowers the conversion efficiency and decreases the highest generated harmonic order. Plasma investigations using the Z-scan technique confirmed the limiting role of processes such as self-defocusing and nonlinear absorption in the conversion of the wavelength of laser probe pulses.

Optimizing HHG in a plasma involves fulfilling several conditions. Because the time of plasma development ranges from several nanoseconds to several hundred nanoseconds, the passage of a laser pulse should be timed to the occurrence of the "optimally prepared" plasma. The investigations described above showed that the main parameters that must be optimized for fulfilling these conditions are the plasma density, the degree of plasma excitation and ionization, and the linear dimensions of the plasma region. The presence of excited components in the plasma may substantially strengthen the nonlinear optical response of the plasma plume. Another important parameter is the delay between the subnanosecond heating pulse and the femtosecond probe pulse. Also noteworthy is the important role of the experimental geometry, which permits production of both a point-like plasma and a plume elongated along the axis of the probe pulse propagation. Increasing the linear dimensions of the plasma region will affect HHG due to the competition between the increase in the HHG efficiency, which is caused by the elongation of the nonlinear medium, and the reabsorption of the generated harmonics along with destructive interference, owing to the fact that the length of the medium exceeds the length of the coherent interaction of the harmonics in different spectral regions. These features were noted in the studies of HHG in indium and boron plasmas [35, 52] and in experiments involving graphite [58], magnesium [59], aluminum [60], molybdenum [61], and lithium [62].

References

1. Gibbon, P. (2005). *Short Pulse Laser Interactions with Matter: An Introduction*, Imperial College Press, London.
2. von der Linde, D. and Rzazewski, K. (1996). High-order optical harmonic generation from solid surfaces, *Appl. Phys. B*, **63**, 499–506.
3. Jaegle, P. (2006). *Coherent Sources of XUV Radiation: Soft X-Ray Lasers and High-Order Harmonic Generation*, Springer, New York.
4. Milošević, D.B. and Ehlotzky, F. (2003). Scattering and reaction processes in powerful laser fields, *Adv. At. Mol. Opt. Phys.*, **49**, 373–532.
5. Antoine, P., L'Huillier, A., Lewenstein, M. *et al.* (1996). Theory of high-order harmonic generation by an elliptically polarized laser field, *Phys. Rev. A*, **53**, 1725–1745.
6. Corkum, P.B. (1993). Plasma perspective on strong field multiphoton ionization, *Phys. Rev. Lett.*, **71**, 1994–1997.
7. Krause, J.L., Schafer, K.J. and Kulander, K.C. (1992). High-order harmonic generation from atoms and ions in the high intensity regime, *Phys. Rev. Lett.*, **68**, 3535–3538.
8. Lewenstein, M., Balcou, P., Ivanov, M.Y. *et al.* (1994). Theory of high-harmonic generation by low-frequency laser fields, *Phys. Rev. A*, **49**, 2117–2132.
9. Miyazaki, K. and Takada, H. (1995). High-order harmonic generation in the tunneling regime, *Phys. Rev. A*, **52**, 3007–3021.
10. Christov, I.P., Murnane, M.M. and Kapteyn, H.C. (1997). High-harmonic generation of attosecond pulses in the "single-cycle" regime, *Phys. Rev. Lett.*, **78**, 1251–1254.

11. Tempea, G., Geissler, M., Schnürer, M. *et al.* (2000). Self-phase-matched high harmonic generation, *Phys. Rev. Lett.*, **84**, 4329–4332.

12. Donnelly, T.D., Ditmire, T., Neuman, T. *et al.* (1996). High-order harmonic generation in atom clusters, *Phys. Rev. Lett.*, **76**, 2472–2475.

13. Hu, S.X. and Xu, Z.Z. (1997). Enhanced harmonic emission from ionized clusters in intense laser pulses, *Appl. Phys. Lett.*, **71**, 2605–2607.

14. Ivanov, M.Y. and Corkum, P.B. (1993). Generation of high-order harmonics from inertially confined molecular ions, *Phys. Rev. A*, **48**, 580–590.

15. Liang, Y., Augst, S., Chin, S.L. *et al.* (1994). High harmonic generation in atomic and diatomic molecular gases using intense picosecond laser pulses: a comparison, *J. Phys. B: At. Mol. Opt. Phys.*, **27**, 5119–5130.

16. Akiyama, Y., Midorikawa, K., Matsunawa, Y. *et al.* (1992). Generation of high-order harmonics using laser-produced rare-gas-like ions, *Phys. Rev. Lett.*, **69**, 2176–2179.

17. Kubodera, S., Nagata, Y., Akiyama, Y. *et al.* (1993). High-order harmonic generation in laser-produced ions, *Phys. Rev. A*, **48**, 4576–4582.

18. Wahlström, C.-G., Borgström, S., Larsson, J. *et al.* (1995). High-order harmonic generation in laser-produced ions using a near-infrared laser, *Phys. Rev. A*, **51**, 585–591.

19. Theobald, W., Wülker, C., Schäfer, F.R. *et al.* (1995). High-order harmonic generation in carbon vapour and low charged plasma, *Opt. Commun.*, **120**, 177–183.

20. Gladkov, S.M. and Koroteev, N.I. (1990). Quasiresonant nonlinear optical processes involving excited and ionized atoms, *Sov. Phys. Usp.*, **33**, 554–575.

21. Ganeev, R.A., Elouga Bom, L.B., Kieffer, J.-C. *et al.* (2007). Optimum plasma conditions for the efficient high-order harmonic generation in platinum plasma, *J. Opt. Soc. Am. B*, **24**, 1319–1323.

22. Kuroda, H., Suzuki, M., Ganeev, R. *et al.* (2005). Advanced 20 TW Ti:S laser system for X-ray laser and coherent XUV generation irradiated by ultra-high intensities, *Laser Part. Beams*, **23**, 203–206.

23. Ganeev, R.A., Naik, P.A., Singhal, H. *et al.* (2007). Tuning of the high-order harmonics generated from laser plasma plumes and solid surfaces by varying the laser spectrum, chirp, and focal position, *J. Opt. Soc. Am. B*, **24**, 1138–1143.

24. Ganeev, R.A., Suzuki, M., Redkin, P.V. *et al.* (2007). Variable pattern of high harmonic spectra from a laser-produced plasma by using the chirped pulses of narrow-bandwidth radiation, *Phys. Rev. A*, **76**, 023832.

25. Ozaki, T., Elouga Bom, L.B., Ganeev, R. *et al.* (2007). Intense harmonic generation from silver ablation, *Laser Part. Beams*, **25**, 321–327.

26. Suzuki, M., Baba, M., Ganeev, R.A. *et al.* (2007). Observation of single harmonic enhancement due to quasi-resonance conditions with the tellurium ion transition in the range of 29.44 nm, *J. Opt. Soc. Am. B*, **24**, 2686–2689.

27. Ganeev, R.A., Elouga Bom, L.B. and Ozaki, T. (2007). High-order harmonic generation from plasma plume pumped by 400-nm wavelength laser, *Appl. Phys. Lett.*, **91**, 131104.

28. Ganeev, R.A., Elouga Bom, L.B. and Ozaki, T. (2007). Enhancement of the high-order harmonic generation from the gold plume using the time-resolved plasma spectroscopy, *J. Appl. Phys.*, **102**, 073105.

29. Ganeev, R.A., Baba, M., Suzuki, M. *et al.* (2005). High-order harmonic generation from silver plasma, *Phys. Lett. A*, **339**, 103–109.

30. Milošević, D.B. (2006). Theoretical analysis of high-order harmonic generation from a coherent superposition of states, *J. Opt. Soc. Am. B*, **23**, 308–317.

31. Ganeev, R.A., Elouga Bom, L.B., Ozaki, T. *et al.* (2007). Maximizing the yield and cutoff of high-order harmonic generation from plasma plume, *J. Opt. Soc. Am. B*, **24**, 2770–2778.

32. Singhal, H., Ganeev, R.A., Naik, P.A. *et al.* (2008). Dependence of high order harmonics intensity on laser focal spot position in pre-formed plasma plumes, *J. Appl. Phys.*, **103**, 013107.

33. Ozaki, T., Elouga Bom, L.B., Ganeev, R.A. *et al.* (2008). Extending the capabilities of ablation harmonics to shorter wavelengths and higher intensity, *Laser Part. Beams*, **26**, 235–240.

34. Ganeev, R., Suzuki, M., Baba, M. *et al.* (2005). High-order harmonic generation from boron plasma in the extreme-ultraviolet range, *Opt. Lett.*, **30**, 768–770.

35. Ganeev, R.A., Baba, M., Suzuki, M. *et al.* (2006). Optimization of harmonic generation from boron plasma, *J. Appl. Phys.*, **99**, 103303.

36. Ganeev, R.A., Suzuki, M., Baba, M. *et al.* (2006). Frequency conversion in laser-produced boron plasma using longitudinal pump scheme, *Eur. Phys. J. D*, **37**, 255–259.

37. Ganeev, R.A., Singhal, H., Naik, P.A. *et al.* (2007). Optimization of the high-order harmonics generated from silver plasma, *Appl. Phys. B*, **87**, 243–247.

38. Kim, H.T., Tosa, V. and Nam, C.H. (2006). Synchronized generation of bright high-order harmonics using self-guided and chirped femtosecond laser pulses, *J. Phys. B: At. Mol. Opt. Phys.*, **39**, S265–S274.

39. Tosa, V., Takahashi, E., Nabekawa, Y. *et al.* (2003). Generation of high-order harmonics in a self-guided beam, *Phys. Rev. A*, **67**, 063817.

40. Brimhall, N., Painter, J.C., Powers, N. *et al.* (2007). Measured laser-beam evolution during high-order harmonic generation in a semi-infinite gas cell, *Opt. Express*, **15**, 1684–1689.

41. Elouga Bom, L.B., Kieffer, J.-C., Ganeev, R.A. *et al.* (2007). Influence of the main pulse and prepulse intensity on high-order harmonic generation in silver plasma ablation, *Phys. Rev. A*, **75**, 033804.

42. Larsen, J.T. and Lane, S.M. (1994). HYADES — A plasma hydrodynamics code for dense plasma studies, *J. Quant. Spectrosc. Radiat. Transf.*, **51**, 179–186.

43. Rubenchik, A.M., Feit, M.D., Perry, M.D. *et al.* (1998). Numerical simulation of ultra-short laser pulse energy deposition and bulk transport for material processing, *Appl. Surf. Sci.*, **129**, 193–198.

44. Wood-Vasey, W.-M., Budil, K.S., Remington, B.A. *et al.* (2000). Computational modeling of classical and ablative Rayleigh–Taylor instabilities, *Laser Part. Beams*, **18**, 583–593.

45. Tillack, M.S. Sequoia, K.L., O'Shay, J. *et al.* (2006). Optimization of plasma uniformity in laser-irradiated underdense targets, *J. Phys. IV*, **133**, 985–988.

46. Ganeev, R.A., Suzuki, M., Baba, M. *et al.* (2005). Generation of strong coherent extreme ultraviolet radiation from the laser plasma produced on the surface of solid targets, *Appl. Phys. B*, **81**, 1081–1089.

47. Reintjes, J.F., She, C.Y. and Eckardt, R. (1978). Generation of coherent radiation in XUV by fifth- and seventh-order frequency conversion in rare gases, *IEEE J. Quantum Electron.*, **14**, 581–596.

48. Pfeifer, T., Gallmann, L. and Abel, M.J. (2006). Single attosecond pulse generation in the multicycle-driver regime by adding a weak second-harmonic field, *Opt. Lett.*, **31**, 975–977.
49. Ishikawa, K.L., Takahashi, E.J. and Midorikawa, K. (2007). Single-attosecond pulse generation using a seed harmonic pulse train, *Phys. Rev. A*, **75**, 021801.
50. Zeng, Z.N. Cheng, Y., Song, X.N. *et al.* (2007). Generation of an extreme ultraviolet supercontinuum in a two-color laser field, *Phys. Rev. Lett.*, **98**, 203901.
51. Ganeev, R.A., Naik, P.A., Singhal, H. *et al.* (2007). Strong enhancement and extinction of single harmonic intensity in the mid- and end-plateau regions of the high harmonics generated in low-excited laser plasmas, *Opt. Lett.*, **32**, 65–67.
52. Ganeev, R.A., Singhal, H., Naik, P.A. *et al.* (2006). Harmonic generation from indium-rich plasmas, *Phys. Rev. A*, **74**, 063824.
53. Ganeev, R.A., Singhal, H., Naik, P.A. *et al.* (2006). Single harmonic enhancement by controlling the chirp of the driving laser pulse during high-order harmonic generation from GaAs plasma, *J. Opt. Soc. Am. B*, **23**, 2535–2540.
54. Ganeev, R.A., Suzuki, M., Baba, M. *et al.* (2006). Analysis of the nonlinear self-interaction of femtosecond pulses during high-order harmonic generation in laser-produced plasma, *J. Opt. Soc. Am. B*, **23**, 1332–1337.
55. Ganeev, R.A., Suzuki, M., Baba, M. *et al.* (2007). High harmonic generation from the laser plasma produced by the pulses of different duration, *Phys. Rev. A*, **76**, 023805.
56. Ganeev, R.A., Elouga Bom, L.B., Kieffer, J.-C. *et al.* (2007). Demonstration of the 101st harmonic generated from laser-produced manganese plasma, *Phys. Rev. A*, **76**, 023831.
57. Tosa, V., Kim, H.T., Kim, I.J. *et al.* (2005). High-order harmonic generation by chirped and self-guided femtosecond laser pulses. II. Time-frequency analysis, *Phys. Rev. A*, **71**, 063808.
58. Ganeev, R.A., Suzuki, M., Baba, M. *et al.* (2005). High-order harmonic generation from carbon plasma, *J. Opt. Soc. Am. B*, **22**, 1927–1933.
59. Ganeev, R.A. and Kuroda, H. (2005). Frequency conversion of femtosecond radiation in magnesium plasma, *Opt. Commun.*, **256**, 242–247.
60. Ganeev, R.A., Baba, M., Suzuki, M. *et al.* (2006). 33rd harmonic generation from aluminum plasma, *J. Modern Opt.*, **53**, 1451–1458.
61. Ganeev, R.A., Kulagin, I.A., Suzuki, M. *et al.* (2005). Harmonic generation in Mo plasma, *Opt. Commun.*, **249**, 569–577.
62. Suzuki, M., Baba, M., Ganeev, R.A. *et al.* (2012). Dependence of the cutoff in lithium plasma harmonics on the delay between the prepulse and the main pulse, *J. Phys. B: At. Mol. Opt. Phys.*, **45**, 065601.

3

Resonance-Induced Enhancement of High-Order Harmonic Generation in Plasma

A bright, monochromatic, coherent source in the XUV range could be useful for many applications. Among the applications one could mention the study of the time evolution of surface states [1], atomic and molecular spectroscopy [2–4], generation of attosecond pulses [5], interferometry and holography [6], nonlinear optics using harmonic generation [6], etc. X-ray lasers and high-order harmonics are the coherent sources of radiation in this region. Although the latter sources have better coherence than X-ray lasers, they are not monochromatic, unless one selects some harmonic using a monochromator. The use of phase matching conditions to replace the well-known plateau of harmonic distribution by a group of intense harmonics is a step in this direction [7, 8]. Though the generation of an intense single harmonic looks unrealistic at the moment, there are some techniques that can considerably improve single-harmonic yield in relation to to the neighboring harmonics. The possibility of the enhancement of HHG in gaseous media using atomic and ionic resonances has been extensively studied theoretically [9, 10]. This approach could be an alternative to phase matching of the pump and harmonic waves using gas-filled waveguides [11]. Furthermore, for the plasmas generated on the surfaces of some solid targets, the resonance conditions between the harmonic wavelength and the excited states of neutrals and singly charged ions can lead to enhancement of the yield for some specific harmonic orders

[12, 13]. The availability of a much wider range of target materials for plasma HHG compared to a few gases increases the possibility of resonance of an ionic transition with a harmonic wavelength in the former case.

At the beginning of a new phase of plasma HHG studies, a considerable variation of the harmonic distribution during HHG from indium plasma was reported [14]. The resonance-induced enhancement was achieved by changing the harmonic wavelength, which enabled the overlap of some specific harmonic with the ionic transition possessing strong oscillator strength. Another approach for the tuning of harmonic wavelength based on chirp variations of driving radiation has been reported during gas HHG studies [15–21]. Moreover, the tuning of harmonics and their "sharpness" were demonstrated using a combination of external control of laser chirp and intensity-induced variation of laser chirp inside a nonlinear medium [16, 17]. In this chapter, we describe studies of the enhancement of single-harmonic intensity in different parts of the plateau in the case of plasma HHG. We analyze giant single-harmonic generation in indium plasma and other plasmas where harmonic enhancement has been reported. We also discuss the conditions of single-harmonic enhancement with strong excitation of the plasma by femtosecond radiation and using a UV laser. The theoretical approaches for explanation of observed single-harmonic enhancement are also analyzed in this section.

3.1. Giant Enhancement of 13th Harmonic Generation in Indium Plasma

The reviewed studies were carried out using the ultrashort pulses (35, 48, and 150 fs) of chirped-pulse amplification Ti:sapphire lasers, operating at a 10 Hz pulse repetition rate. The details of the experimental arrangements are analogous to those described previously and can be found elsewhere [22, 23]. The absolute calibration of harmonic efficiency was performed using the technique described in [24].

The first demonstration of giant resonance-induced enhancement of the single harmonic generated in plasma was reported in [14]. A typical harmonic spectrum obtained from indium plasma using 796 nm, 150 fs laser pulses is shown in Fig. 3.1. High-order harmonics were observed in these experiments and showed a plateau pattern. The conversion efficiency at the

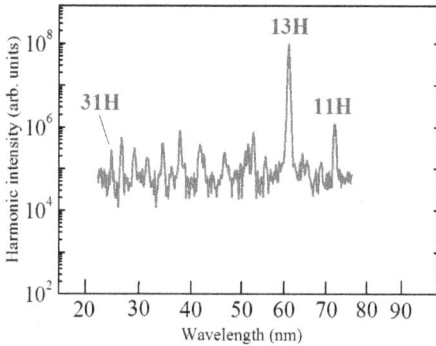

Fig. 3.1 High-order harmonic spectrum generated in indium plasma. Reprinted from [14] with permission from Optical Society of America.

plateau region in the case of indium plasma was measured to be 8×10^{-7}. The most intriguing feature observed in these studies was a very strong 13th harmonic, whose intensity was almost two orders of magnitude higher than those of the neighboring harmonics. The conversion efficiency of the 13th harmonic was 8×10^{-5}, and for the pump energy of 10 mJ, this corresponded to 0.8 μJ [14].

After the observation of such an unusual harmonic distribution, the question arises whether the strong emission associated with the 13th harmonic ($\lambda = 61.2$ nm) originates from amplified spontaneous emission, re-excitation of plasma by a femtosecond beam, or a nonlinear optical process related to the enhancement of a single harmonic due to its spectral proximity to the resonance transitions. The probe laser polarization was varied to analyze the strong emission near 61 nm in the harmonic spectrum. A small deviation from linear polarization led to a considerable decrease of the 61.2 nm radiation intensity, which is typical behavior for the high-order harmonics. The application of circularly polarized laser pulses led to the complete disappearance of 61.2 nm emission. At the same time, the excited lines of the plasma spectrum observed at different polarizations of the probe beam remained unchanged, which clearly shows that the strong 61.2 nm emission was generated through the nonlinear optical process.

The wavelength of the probe beam was tuned to analyze whether the excited ionic transitions from indium plasma influence the plateau pattern of the harmonic distribution. The central wavelength of the output radiation of a Ti:sapphire laser was tuned between 770 and 796 nm. The 13th harmonic output was considerably decreased with the detuning of the fundamental wavelength from 796 nm toward the shorter wavelength region. At the same time, a strong enhancement of the 15th harmonic in the case of the 782 nm radiation was observed, while the intensities of other harmonics remained relatively unchanged. These observations have shown the influence of ionic transitions on the intensity of individual harmonics. In particular, it was demonstrated that the 26 nm shift of the central wavelength of the probe pulse, which corresponded to the 2 nm shift of the wavelength of the 13th harmonic, considerably changed the overall pattern of the harmonic distribution at the plateau region.

The observed phenomenon of giant enhancement of the single harmonic in indium plasma has recently been carefully analyzed [23, 25]. These studies have shown that the application of shorter pulses (35 and 48 fs) can further improve the output of the 13th harmonic from indium plasma. A 10^{-4} conversion efficiency to the 61 nm line has been reported, with a 200 times enhancement factor for the 13th harmonic with regard to the neighboring ones.

The question arises of why one can achieve a strong enhancement of the single harmonic within the harmonic spectrum in the case of indium plasma. A comparison with a past study of indium plasma [26] shows that the emission in the range of 40–65 nm is due to radiative transitions to the ground state ($4d^{10}$ $5s^2$ $^1S_0$) and low-lying state ($4d^{10}$ 5s 5p) of In II. Previous work [26] revealed an exceptionally strong line at 62.1 nm (19.92 eV), corresponding to the $4d^{10}$ $5s^2$ $^1S_0$ → $4d^9$ $5s^2$ 5p (^2D) $^1P_1$ transition of In II. The oscillator strength gf of this transition has been calculated to be 1.11, which is more than 12 times larger than those of other transitions in this spectral range. This transition can be driven into resonance with the 13th harmonic ($\lambda = 61.2$ nm, $E_{ph} = 20.26$ eV) by the AC Stark shift, thereby resonantly enhancing its intensity. Such intensity enhancement can be attributed to the existence of oscillating electron trajectories that revisit the ionic core twice per laser cycle [10]. Since such trajectories start from the resonantly populated excited state, with a nonzero initial kinetic energy, they still have nonzero instantaneous kinetic energies when they return to the origin. As usual, recombination results in the emission of harmonics, but due to the relatively low probabilities, the population in the laser-driven wavepackets increases continuously and the probability for harmonic emission grows with the number of allowed recollisions. This multiple recollision is predicted to enhance harmonics in the spectral ranges close to the atomic and ionic resonances.

3.2. Single Harmonic Enhancement in Chromium, Gallium Arsenide, and Indium Antimonide Plasmas

The tuning of harmonic wavelength by adjustment of the master oscillator of a laser is not an easily performed process. As mentioned above, a much simpler way of tuning the harmonic wavelength without tuning the laser spectrum can be performed by controlling the chirp of the driving radiation [15, 17, 27]. The laser chirp conditions should be correctly chosen to suppress the harmonic chirp that broadens the high-order harmonics and reduces their peak intensities. In order to analyze the harmonic variations at different chirps of driving radiation after the optimization of various experimental conditions for HHG, further studies have been carried out, which were related to the change of the spectral distribution within the driving and harmonic pulses. The chirp of a driving laser pulse can be varied by adjusting the separation of the gratings in the pulse compressor. A reduction in the grating separation from the chirp-free condition generates positively chirped pulses, and an increase of the grating separation provides negatively chirped pulses.

High-order harmonic generation from the plasmas of a number of targets was studied prior to choosing the samples that are suitable for the observation of single-harmonic enhancement in the short-wavelength range. The harmonic distribution from most of the plasmas showed a featureless plateau-like shape in the XUV range, while in some cases it demonstrated a steady or even steep decrease of conversion efficiency for each successive harmonic order. The harmonics of 48 fs, 795 nm radiation generated from chromium, gallium arsenide (GaAs), and indium antimonide (InSb) plasmas were studied in more detail due to the observation of an abnormal harmonic distribution in the plateau region. These observations were compared with the harmonics generated in indium plasma.

Harmonics up to the 29th (27.3 nm), 43rd (18.4 nm), and 47th (16.9 nm) orders were observed in these experiments with InSb, GaAs, and chromium plasmas, respectively [18]. These targets showed some enhancement (or decrease) of specific harmonic order. The studies were performed using

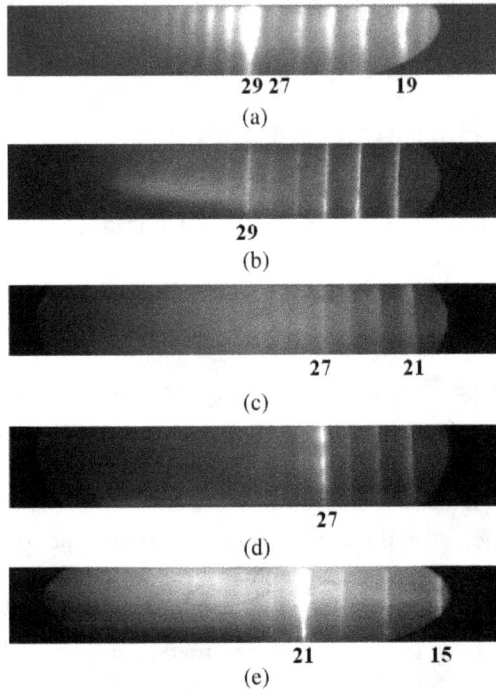

Fig. 3.2 Harmonic distribution in the case of a chromium plume at different chirps of the driving pulses: (a) chirp-free 48 fs pulses; (b) negatively chirped 160 fs pulses. Harmonic distribution in the case of GaAs plume: (c) chirp-free 48 fs pulses; (d) positively chirped 130 fs pulses. (e) Harmonic distribution in the case of InSb plasma and chirp-free 48 fs driving pulses. Adapted from [18] with permission from Optical Society of America.

chirp-free 48 fs pulses, as well as chirped pulses. The change of laser chirp resulted in a considerable variation of harmonic distribution in the mid- and end-plateau ranges, due to the tuning of some harmonics toward the ionic transitions [27]. Figure 3.2 shows the spectra of harmonics generated from those plasmas.

The HHG in chromium plasma showed a considerable variation of the 27th harmonic intensity at different chirps of the driving radiation [28]. At some chirps, the 27th harmonic almost disappeared from the harmonic spectrum. At the same time, a strong 29th harmonic (27.3 nm) was observed in the case of chirp-free pulses. The chirp variation led to a change in the 29th harmonic yield compared with the neighboring harmonics. The maximum ratio of the 31st and 29th harmonic intensities was measured to be 23. It should be noted that for negatively chirped pulses the harmonics became narrower. With further growth of negative chirp the harmonic spectrum was considerably detuned from the resonances causing the absorption of the 27th harmonic and the enhancement of the 29th harmonic. In that case, the intensities of these harmonics became comparable with each other [28].

Analogous variations of the harmonic distribution at the end- and mid-plateau ranges were observed in the cases of GaAs and InSb plumes [23, 29]. The change of laser chirp resulted in tuning of the harmonics generated in the GaAs plasma. At the chirp-free case and for negatively chirped pulses, a feature-less plateau-like distribution of high-order harmonics with a gradual decrease of harmonic intensity was observed. However, for positively chirped pulses, an enhanced 27th harmonic (29.4 nm) appeared. The intensity of this harmonic was six times higher than the intensities of neighboring lower-order harmonics.

In the case of the InSb plume, a strong 21st harmonic (37.8 nm) of chirp-free driving radiation was observed. These studies showed an eight-fold enhancement of the 21st harmonic compared with the neighboring harmonics. The 21st harmonic intensity varied at different chirps of the pump laser. In particular, in the case of positively chirped 140 fs pulses, the 21st harmonic exceeded the neighboring ones by a factor of ten. The enhancement of this harmonic considerably decreased in the case of negatively chirped pulses [13].

The origin of the enhanced emission in the vicinity of 27.3 nm (the 29th harmonic from the chromium plasma), 29.4 nm (the 27th harmonic from the GaAs plasma), and 37.8 nm (the 21st harmonic from the InSb plasma) was analyzed by inserting a quarter-wave plate in the path of the femtosecond laser beam. No harmonics appeared for circularly polarized laser pulses, as expected by assuming the nonlinear optical origin of the observed spectra.

The enhancement of the single harmonics belonging to the mid- and end-plateau regions was smaller than the enhancement of the single harmonic generated in the indium plasma. The enhancement of the 13th harmonic (200×) generated in the indium plasma considerably exceeded the enhancements for the 21st (10×), 27th (6×), and 29th (23×) harmonics generated in the InSb, GaAs, and chromium plasmas, respectively. Such a difference was attributed to the different oscillator strengths of the ionic transitions involved in the resonance enhancement of harmonics.

It should be mentioned that the resonant enhancement of single high-order harmonics discussed here has a considerably different origin from the resonant-like enhancement of the groups of high-order harmonics, which has been theoretically demonstrated at particular laser intensities in gaseous media [9, 10, 30]. At present, there is no consensus about the physical origin of these resonant-like enhancements. One group of approaches explains them as laser-intensity-induced channel closings, which are related to long quantum orbits [30]. The second group of explanations suggests that the enhancement occurs at particular intensities where the atom (ion) is in a state that is a superposition of the laser-dressed ground state and of one or several laser-dressed excited states (or laser-induced states) [10]. Recently developed explanations of resonant enhancement of single harmonics are reviewed in Section 3.6.

It follows from the preceding results that the origin of the strong yield of single harmonics in the plateau region is associated with the resonance-induced growth of nonlinear optical process. Therefore, let us examine the resonance-induced growth mechanism in a little more detail. Some experimental observations (in particular, the dependences of the harmonic yield on the beam waist position, plasma size, and laser radiation intensity) show effects related to the collective character of HHG from laser plasma. Among the factors enhancing harmonic output are effects related to the difference in the phase conditions for different harmonics. The phase mismatch ($\Delta k = nk_1 - k_i$, where k_1 and k_i are the wave numbers of the probe radiation and the ith harmonic) changes due to the ionization caused by propagation of the driving pulse through the plasma. According to calculations [31], the phase mismatch caused by the influence of free electrons is about one to two orders of magnitude higher than those caused by the influence of atoms and ions. However, at resonance conditions, when the harmonic frequency is close to the frequency of an atomic transition, the variation of the wave number of a single harmonic could be considerable, and the influence of free-electron-induced mismatch can be compensated by the atomic dispersion for a specific harmonic order. In that case, improvement of the phase matching conditions for single-harmonic generation can be achieved. Such a mechanism may be partially responsible for the enhancement of the above-described nonlinear optical processes in indium, GaAs, chromium, and InSb plasmas. Analogous single harmonic enhancement has recently been reported in tin [12] and antimony [13] plasmas, while in the case of manganese plasma, multiple harmonic enhancement was observed [32].

3.3. Single Harmonic Enhancement at Strong Excitation Conditions

In the above described studies, no significant difference between the sharpness of harmonics in different parts of the plateau region was observed in the case of chirp-free laser pulses. No considerable influence of self-phase modulation (SPM) on the spectral distribution of harmonics is expected, since the experimental conditions (low-density plasma, moderate laser intensities) restricted the possibility of the influence of the strongly ionized medium on the phase characteristics of harmonics. An over-ionized medium, with electron density in the central part higher than in the outer region, acts as a negative lens, leading to a defocusing of the laser beam in a plasma and hence to a reduction in the effective harmonic generation volume. In addition, the rapidly ionizing high-density medium modifies the temporal structure of the femtosecond laser pulse due to SPM.

The studies of HHG in silver plasma were carried out to demonstrate harmonic tuning at variable intensities of the driving radiation, which allowed in some cases achievement of the resonance enhancement of single harmonics [27]. The maximum intensity of the femtosecond beam at the focal spot was 4×10^{17} W cm^{-2}. This intensity considerably exceeded the barrier suppression intensity for singly charged silver ions. Necessary care was taken to optimize the high-order harmonic output by adjusting the position of the laser focus by placing it in front of the laser plasma. The intensity of the driving laser pulse at the preformed plasma in that case was varied between 1×10^{14} and 5×10^{15} W cm^{-2}. This intensity was varied by changing the position of the focusing lens with respect to the position of the plasma. It was shown that, as the focal spot approaches the plasma area, which corresponds to the increase of laser intensity inside the plasma, the harmonics were tuned up to 1.2 nm (for the 19th harmonic generated in the silver plasma).

Tuning of the harmonic wavelength could be achieved by (i) tuning the fundamental wavelength of the laser pulse [14, 33], (ii) chirping the laser radiation [15, 18, 25], (iii) altering the laser intensity, which leads to the control of the ionization rate of the nonlinear medium [27, 34, 35], and (iv) adaptive control using shaped laser pulses [36]. There is also another method of harmonic tuning. Second-harmonic generation from nonlinear optical crystals (β-barium borate (BBO) or potassium dihydrogen phosphate (KDP)) can be optimized at different phase matching conditions, thus allowing tuning of the second-harmonic wavelength. This is followed by tuning of the high-order harmonics using second-harmonic pulses as the driving radiation.

Note that effective tuning using the chirping technique and the variable second-harmonic radiation technique can be achieved only in the case of broadband radiation. In that case, only the leading part of the fundamental pulse consisting of either blue or red components participates in HHG. In the case of narrowband radiation, the difference in the components at the leading and trailing parts of the chirped pulse is negligible. However, the interplay between the SPM-induced chirp and the artificial chirp of narrowband radiation induced by variation of the distance between the gratings in the compressor can lead to the adjustment of some harmonics toward quasi-resonance conditions, which in turn can alter the harmonic output. This process has been reported during HHG studies in the case of a plasma containing gallium nitride (GaN) nanoparticles [19].

A GaN powder consisting of 20 nm nanoparticles was glued onto the substrates and then ablated by a heating pulse. Figure 3.3 presents a few lineouts of the harmonic spectra at different chirps of the narrowband driving radiation (10 nm full width at half maximum). The harmonic generation was followed by plasma emission from GaN nanoparticles in the XUV region,

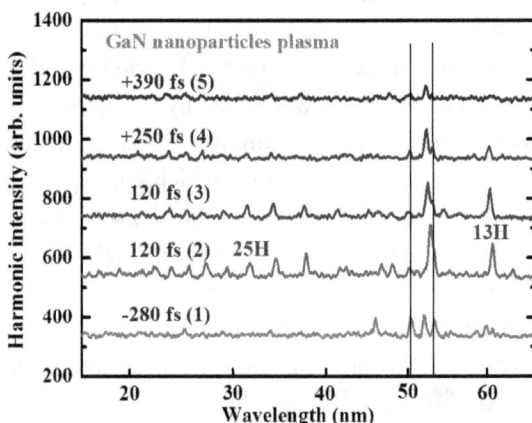

Fig. 3.3 Harmonic spectra obtained in GaN nanoparticle plasma using radiation of different chirps and pulse durations. Reprinted from [19] with permission from American Physical Society.

which was dominated by the strong 50.3 and 53.8 nm ionic transitions marked by two lines. In the case of negatively chirped 280 fs pulses, the 15th harmonic wavelength (52.7 nm) was just between these two ionic transitions (curve 1). The application of chirp-free pulses led to the redshift-induced tuning of the 15th harmonic toward the longer-wavelength transition (curve 2). This redshift is consistent with a previously reported discussion of the role of nanoparticles and clusters in the red shift of fundamental radiation wavelength [37]. Simultaneously, both the enhanced 15th harmonic and higher-order harmonics were the main features of this spectral pattern. A decrease of the intensity of chirp-free pulses (by decreasing the pulse energy) led to less influence of the SPM-induced redshift, which caused the detuning of this harmonic toward the blue side, where one can distinguish the ionic transition at the longer-wavelength wing of 15th harmonic spectrum (curve 3). The positive chirp led to further detuning of this harmonic toward the shorter-wavelength ionic transition due to the decrease of pulse intensity (curves 4 and 5).

The experiments with GaN nanoparticle-containing plasma revealed the opportunity for resonance-induced modification of harmonic spectra using an intensity-induced shift of harmonic wavelength. More details on this approach are described in the following section. As discussed, the resonance-induced enhancement of the single harmonic was achieved using chirping of the broadband laser pulses. This led to the tuning of harmonics toward the strong ionic transitions due to re-distribution of the laser spectrum along the main pulse and preferential harmonic generation for the leading part of this pulse [18, 29]. The defocusing properties of plasma weakened the trailing part of the pulse. That is why the leading part of the chirped pulse was mainly responsible for the harmonic tuning.

Control over tuning through a change of average intensity leading to variation of the ionization state of an argon-filled capillary has recently been examined [35]. The harmonic tuning was shown to depend on the nonlinear spectral shift of the fundamental laser pulse caused by ionization, rather than directly on the artificial chirp imposed on the fundamental wavelength. In most resonance-related HHG experiments, the harmonic spectrum was studied as a function of laser intensity to show the existence of enhancements for particular intensities [10, 38, 39]. This technique is considerably different from the early approach where the variation of the driving radiation spectrum tuned the harmonic wavelength and adjusted it to coincide with the atomic or ionic resonances in the nonlinear medium. The above-discussed approach is close to the latter one, though no change of the spectrum of the fundamental radiation was imposed; instead the change of the spectral distribution inside the pulse was carried out by controlling the chirp of the laser radiation.

3.4. Resonance Enhancement of Odd and Even Harmonics in Tin Plasma During Two-Color Pumping

As mentioned above, the enhancement of HHG in gaseous media using atomic and ionic resonances has extensively been studied theoretically, starting over ten years ago [10, 38, 40–42]. Recent proposals to explain the experimentally observed peculiarities of resonant harmonics from laser plasma [43–50] have considerably improved our understanding of the mechanisms governing this process. Those theoretical approaches will be discussed in Section 3.6.

Reviews of early resonance-induced enhanced harmonic studies for various metal and semiconductor targets ablated by picosecond pulses are presented in [51–53]. In those studies, a single pump (800 nm) was used for harmonic generation. It is of interest to analyze the availability of resonance enhancement in plasmas in the case of the two-color pump (800 and 400 nm), since this can give an opportunity for further overlap between some harmonic orders (this time both odd and even) with ionic resonances possessing strong oscillator strengths. Two-color pumping using the fundamental and second-harmonic (SH) fields has become a practical way of harmonic enhancement in gas media [54–59]. The same approach has recently been reported for plasma harmonics when enhanced HHG in laser-produced plasma plumes was observed using a two-color laser pump (98% of 800 nm and 2% of 400 nm [60] and 92% of 800 nm and 8% of 400 nm [61]). Those studies are analyzed in Chapter 6. In the meantime, the use of a plasma plume as the HHG medium in the two-color pump scheme allows the search for appropriate targets where resonance enhancement can be achieved for the even harmonics of a Ti:sapphire laser, analogous to previously reported observations of resonantly enhanced odd harmonics in the case of single-pump schemes [51, 52].

Below, we discuss a study showing that the use of plasma production on tin or tin-containing alloy allows one to obtain the enhanced 18th harmonic of a Ti:sapphire laser, possibly through the AC Stark shift-induced tuning of ionic transitions towards the harmonic wavelength or due to creation of intensity-dependent phase matching conditions for the single harmonic [62]. Harmonic generation was studied in tin and tin alloy ablated by 210 ps pulses and probed by 800 nm and 400 nm, 48 fs pulses. For SH generation, a BBO crystal (1 mm thick, type I phase matched) was inserted between the focusing lens and the plasma plume, so that, after frequency up-conversion in the crystal, the emerging laser field consisted of both the SH and fundamental laser pulses (see inset in Fig. 3.4). The SH conversion efficiency at these conditions was

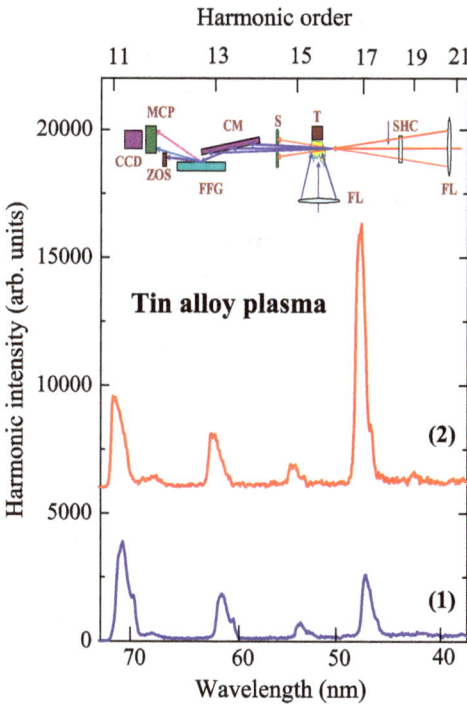

Fig. 3.4 Harmonic spectra from a tin alloy plasma plume in the case of a single-color pump for (1) chirp-free 48 fs pump pulses, and (2) positively chirped 150 fs pump pulses. The intensities of the harmonics are normalized with respect to the intensity of the 11th harmonic. Inset: Experimental scheme. FL, focusing lenses; SHC, second-harmonic crystal; T, target; S, slit; CM, cylindrical mirror; FFG, flat-field grating; ZOS, zero-order beam stop; MCP, microchannel plate; CCD, charge coupled device. Reprinted from [62] with permission from Optical Society of America.

measured to be 8%. The polarizations of the SH and the fundamental fields were orthogonal to each other.

Curve 1 of Fig. 3.4 shows the HHG spectrum from a tin alloy plume, in the wavelength range of 40–75 nm. This pattern was observed in the case of chirp-free 800 nm, 48 fs pulses. One can see the 17th harmonic intensity enhanced compared to that of the lower harmonics. Next, the chirp of the laser pulse was varied by adjusting the separation between the gratings in the pulse compressor. The variation of the chirp of the pump laser towards the positively chirped pulses led to a decrease in the intensity of the low-order harmonics, while the 17th harmonic became stronger (Fig. 3.4, curve 2). In that case, 150 fs positively chirped pulses were used. The maximum ratio of the intensity of the 17th harmonic compared to those of the 11th, 13th, 15th and 19th harmonics was measured to be approximately 3, 4, 11, and 20, respectively. The Sn II ion (I_i = 14.6 eV) has the oscillator strength value 1.52 for the transitions $^2P_{3/2} \rightarrow (^1D)\,^2D_{5/2}$ for which $\Delta\varepsilon$ = 26.27 eV. The laser wavelength that corresponds to the resonance of $\Delta\varepsilon$ with the 17th harmonic is 802 nm. The resonance enhancement was obtained when the wavelengths of the harmonic and the red part of the strong radiative transition were adequately matched. At these conditions, the optimal phase matching and absorption conditions led to the strongest single harmonic generation. Neither copper nor nickel (the other major constituents of the tin alloy) plasma plumes showed the generation of resonantly enhanced harmonics, as checked in separate experiments with pure copper and nickel targets.

Insertion of a nonlinear crystal led to the appearance of a second wave, which enables one to study the joint action of two orthogonally polarized waves during harmonic generation in the plasma plume. In spite of the influence of group velocity mismatch and walk-off of the fundamental and the SH radiation in the relatively thick (1 mm) BBO crystal, calculations show that the ordinary 800 nm wave and extraordinary 400 nm wave still have a sufficient temporal overlap, especially in the case of chirped pulses. The SH crystal was inserted in the experimental scheme at an appropriate position after the focusing lens, so that no impeding processes were observed after propagation of the laser radiation through the crystal. The enhanced odd and even harmonic generation was achieved despite a significant difference between the two pump intensities (12:1), and without any precise temporal and spatial matching of the two pumps. In [61], a 1 mm thick KDP crystal was used for two-color pumping of plasma, which showed better conditions of temporal overlapping of the fundamental and SH waves after leaving the crystal compared with the BBO crystal. However, the conversion efficiency in KDP was considerably less (2%) due to smaller nonlinear susceptibility and narrower angular and spectral phase matching of this crystal. Even at a large difference between the two pump intensities (50:1), the advantages of the two-color pump scheme over the single-color pump were clearly seen by analysis of harmonic enhancement.

The closeness of the wavelengths of the 17th harmonic of the 800 nm pump (47 nm, $E_{ph} = 26.45$ eV) and the 9th harmonic of the 400 nm pump (44.4 nm, $E_{ph} = 28.1$ eV) and strong transitions of singly charged tin ions (25–27 eV) showed identical resonance-induced enhancement [12]. The influence of the Sn III $4d^{10} 5s 5p \rightarrow 4d^9 5s 5p^2$ transitions (28.33–28.71 eV) on the enhancement of the 9th harmonic of 400 nm radiation could be even stronger compared to the Sn II transitions, due to the higher oscillator strengths of the former transitions. Some of these transitions can possibly be driven into resonance with the 18th harmonic by the AC Stark-induced red shift of ionic transition wavelength, thereby resonantly enhancing harmonic intensity. It is difficult to calculate the AC Stark shift accurately, but according to [63], the shifts can be several electronvolts in magnitude. So, the AC Stark shift of these transitions, induced by the strong laser field inside the plasma plume, can in principle be a driving mechanism that drastically changes the relative intensities of the enhanced 17th and 18th harmonics in a two-color pump scheme. Another possible mechanism that can change the harmonic yield with intensity is phase matching. However, this assumption cannot be valid if one compares identical conditions for the neighboring harmonics since, due to small atomic or ionic dispersion of the low-density plasma, the phase matching conditions should be fulfilled for few neighboring harmonics simultaneously.

Figure 3.5 shows the variation of harmonic spectra at different focusing conditions of laser radiation for the tin alloy plasma area. When the two-color laser radiation was focused 4 mm before the plasma plume, a strong 17th harmonic was observed in the spectral range below 50 nm (curve 1). Moving the focus of the laser radiation towards the plasma plume led to the appearance of a strong 18th harmonic (curve 2), alongside the 17th harmonic. Further change of the position of the focal plane towards the plume caused a considerable decrease of the 17th harmonic relative to the 18th harmonic

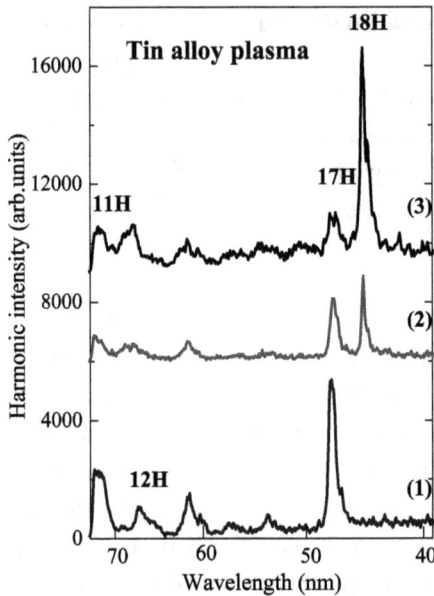

Fig. 3.5 Resonance-induced enhancement of odd (17th) and even (18th) harmonics in the case of tin-containing alloy at different focusing conditions. These curves were taken at (1) 2×10^{14} W cm^{-2} (in the case of focusing 3 mm before the plasma), (2) 5×10^{14} W cm^{-2} (in the case of focusing 2 mm before the plasma), and (3) 2×10^{15} W cm^{-2} (in the case of focusing inside the plasma). Reprinted from [62] with permission from Optical Society of America.

yield (curve 3). At these conditions, the two-color pump was focused inside the plasma plume. Approximately a six-fold enhancement of the 18th harmonic in comparison with neighboring harmonics was achieved by optimizing the position of the focal plane of the two-color pump relative to the plasma plume.

To compare these results attained using tin alloy plasma with those obtained in the plasma produced on the surface of pure tin, analogous studies were carried out with the latter target. The conditions of focusing inside the tin plasma were exceptionally favorable for resonance-enhanced 18th harmonic generation (Fig. 3.6, curve 1). Its intensity almost ten times exceeded that of the other harmonics in the studied spectral range (40–80 nm). One can see the coincidence in the behavior of the harmonic spectra obtained using tin and tin alloy plumes.

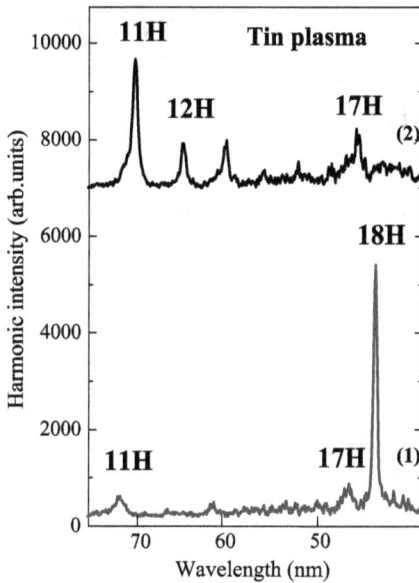

Fig. 3.6 Resonance-induced enhancement of odd (17th) and even (18th) harmonics in the case of pure tin plasma. These curves were taken using (1) chirp-free 48 fs pulses and (2) negatively chirped 160 fs pulses. Reprinted from [62] with permission from Optical Society of America.

These results were obtained using 48 fs, 800 nm chirp-free laser pulses. Variation of the chirp of fundamental radiation at these conditions led to drastic changes in the harmonic spectra. In particular, in the case of negatively chirped 160 fs pulses, lower odd and even harmonics were observed to be stronger than in the case of chirp-free pulses due to a decrease of impeding processes of plasma over-ionization by intense $(1.5 \times 10^{15}$ W cm$^{-2})$ laser pulses in the latter case, together with the moderately enhanced 17th harmonic (Fig. 3.6, curve 2). One can see the blue shift of harmonics compared to the chirp-free case. The most important feature of this spectrum is the disappearance of the 18th harmonic from the harmonic spectra at these conditions, which can be explained by a decrease of pump intensity for chirped pulses. In that case, the role of the AC Stark shift in tuning of the ionic transitions and change in the phase matching conditions becomes insignificant, while the blue shift of 18th harmonic toward the ionic transitions does not compensate for a decrease of the intensity-dependent red shift of Sn II transitions.

3.5. Plasma Harmonic Enhancement Using Two-Color Pump and Chirp Variation of 1 kHz Ti:sapphire Laser

An increase of the pulse repetition rate can ultimately enhance the average power of generated harmonics, thus allowing the improvement of the output characteristics of HHG. Previous studies of resonance-induced and two-color pump-induced enhancement of harmonics in laser plasma were demonstrated using relatively low (10 Hz) pulse repetition rate laser sources [14, 61]. The application of 1 kHz lasers considerably improves the average power of converted laser radiation [64]. More details on applications of high pulse repetition rate lasers for plasma HHG are presented in Chapters 6 and 7.

Resonance enhancement introduces the new possibility of increasing the conversion efficiency of a specific harmonic by more than an order of magnitude [16]. If this effect could be combined with the two-color pumping technique and a high pulse repetition rate, further improvement of the characteristics of HHG would be achieved. The resulting source would have more spectral components due to odd and even harmonics generation, the possibility of high (of order of kHz) repetition rate, and improved conversion efficiency of resonantly enhanced XUV photons. Such a unique radiation source would be ideal for various applications in physics, chemistry, and biology, and for exploring new fields such as nonlinear X-ray optics and attosecond physics. In this section, we describe research when the three above-mentioned advantages — resonance enhancement, two-color pumping, and high pulse repetition rate — were combined to generate strong harmonics in different spectral ranges.

3.5.1. Experimental

A 1 kHz chirped pulse amplification Ti:sapphire laser source was used in these studies. Part of the uncompressed radiation (790 nm, 1.5 mJ, 20 ps, 1 kHz) was split from the beam line prior to the laser compressor stage and was focused into the vacuum chamber to create a plasma on the target surfaces (silver, chromium, vanadium). This picosecond radiation created a plasma plume with a diameter of ~0.4 mm using an intensity on the target surface of $I_{pp} = 5 \times 10^9 - 3 \times 10^{10}$ W cm^{-2}. Broadband femtosecond pulses (780 nm,

30 fs, 1 mJ, 1 kHz, 40 nm spectral width at half maximum) were focused in a direction orthogonal to that of the heating picosecond pulse into the laser plasma using a 200 mm focal length reflective mirror. A 500 mm focal length lens was also used in some experiments instead of the focusing mirror in order to achieve a more extended confocal parameter of the radiation focused inside the plasma area. The position of the focus with respect to the plasma area was chosen to maximize the harmonic signal. The intensity of femtosecond radiation at the focal range was estimated to be $I_{fp} = 5 \times 10^{14}$ W cm^{-2}. The delay between plasma initiation and femtosecond pulse propagation was varied in the range of 6–57 ns using an optical delay line. The harmonic radiation was analyzed by an XUV spectrometer.

The insertion of a nonlinear crystal (BBO) in the path of the 780 nm laser beam led to the appearance of a second wave, 2ω, which enabled the joint action of two orthogonally polarized waves (ω and 2ω) to be studied during harmonic generation in the plasma plume. In spite of the influence of group-velocity mismatch and walk-off of the fundamental and the second-harmonic radiation in the 1 mm thick BBO crystal, calculations show that the ordinary 780 nm wave and extraordinary 390 nm wave still have some temporal overlap. The second-harmonic conversion efficiency did not exceed 4%. Enhanced odd and even harmonic generation was achieved despite a significant difference between the two pump intensities (25:1), and incomplete temporal overlap of the two pulses in the plasma plume. By estimation, there is around a 100 : 1 photon ratio in the plasma area for 780 and 390 nm photons, assuming that duration of the second-harmonic pulse became four times longer after propagation of the 30 fs pulse through the 1 mm thick BBO crystal.

The tuning of harmonic wavelengths was accomplished by using the chirp-induced variation of the distribution of spectral components along the laser pulse. The chirp was varied by adjusting the separation between the two gratings of the pulse compressor. The artificially induced chirp was calibrated by analyzing the phase and pulse duration of laser pulses propagating through the compressor stage at different distances between gratings using the spectral phase interferometry for direct electric-field reconstruction (SPIDER). At these conditions, no new frequencies are introduced, but the "ionization gating" changes the effective drive-laser wavelength in the medium through chirping.

Enhanced harmonic generation in a few plasmas has been analyzed. Below we discuss both single- and two-color pump-induced HHG studies in silver, chromium, and vanadium, when chirp-induced tuning of harmonics allowed the resonance enhancement of single odd and, in some cases, even harmonics, while operating at 1 kHz pulse repetition rate [65].

3.5.2. Silver plasma

Figures 3.7a and 3.7b show two rough images of the parts of the harmonic spectra generated in silver plasma using 780 nm probe radiation. Harmonics above the 50th order were routinely generated. Variation of laser chirp allowed a considerable tuning of harmonic wavelengths (Fig. 3.7c), while the relative intensities of harmonics in the plateau region remained approximately the same over a broad range of spectral phase modulation (from −97 fs to +110 fs; the sign of the pulse duration denotes negative or positive chirp of laser radiation).

The wavelength shift of the harmonics can be explained by the redistribution of spectral components in the leading edge of the chirped radiation. As the intensity of the radiation increases at the leading edge, HHG efficiency is also increased. However, ionization also occurs with the increase of laser intensity, which eventually destroys HHG. Thus there is an optimum femtosecond probe pulse intensity at which the ionization level is still low, but the intensity is still high enough to generate harmonics. This optimum intensity is reached at a specific time within the pulse, and so for chirped pulses there is a specific spectral component associated with this optimum intensity. Therefore, for chirped pulses, the harmonics are odd orders of this spectral component at the leading edge of the pulse. The harmonics produced with positively chirped laser pulses were redshifted because the harmonics produced in the leading edge of the laser pulse come from the red component of the laser spectrum. The same can be said about the blueshifted harmonics produced by negatively chirped pulses. Note that effective tuning can be achieved only in the case of broadband radiation. In that case only the leading part of the fundamental pulse consisting of either blue or red components participates in the HHG. It is worth noting that broadband radiation is associated with a relatively short laser pulse. In the case of narrowband radiation, the difference in the components at the leading and trailing parts of the chirped pulse is not so pronounced. In the experiments described, relatively broadband radiation (40 nm) was used,

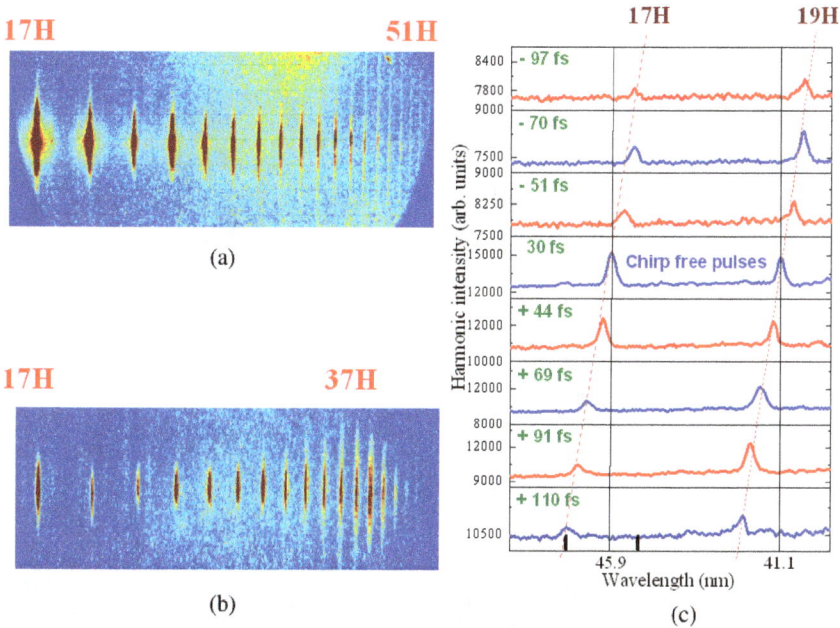

Fig. 3.7 Harmonic spectra from silver plasma in the cases of (a) apertureless and (b) apertured single-color pumping (780 nm). (c) Tuning of the 17th and 19th harmonics by changing the distance between the gratings in the compressor stage. Positive and negative values of pulse duration correspond to positively and negatively chirped pulses. Dotted lines show the tuning of the 17th and 19th harmonics with different chirps. Black lines show the wavelengths of these harmonics at chirp-free conditions. Thick black lines on the left side of the bottom graph show the tuning range of the 17th harmonic (2.8 nm). Reprinted from [65] with permission from Optical Society of America.

which allowed a considerable change in harmonic wavelength with variation of laser chirp, which is crucial for achieving the appropriate adjustment between the wavelengths of some harmonics and ionic resonances possessing strong oscillator strengths.

The introduction of a second orthogonally polarized SH field increased the conversion efficiency of odd harmonics, which has been reported both in gaseous [60] and in plasma [61] HHG studies. In the case of silver plasma, strong even harmonics were observed as well, which were of the same intensity as the odd ones (Fig. 3.8a), and in some cases exceeded them in the cutoff region (Fig. 3.8b), in spite of a considerable difference in the number of interacting ω and 2ω photons in the plasma volume.

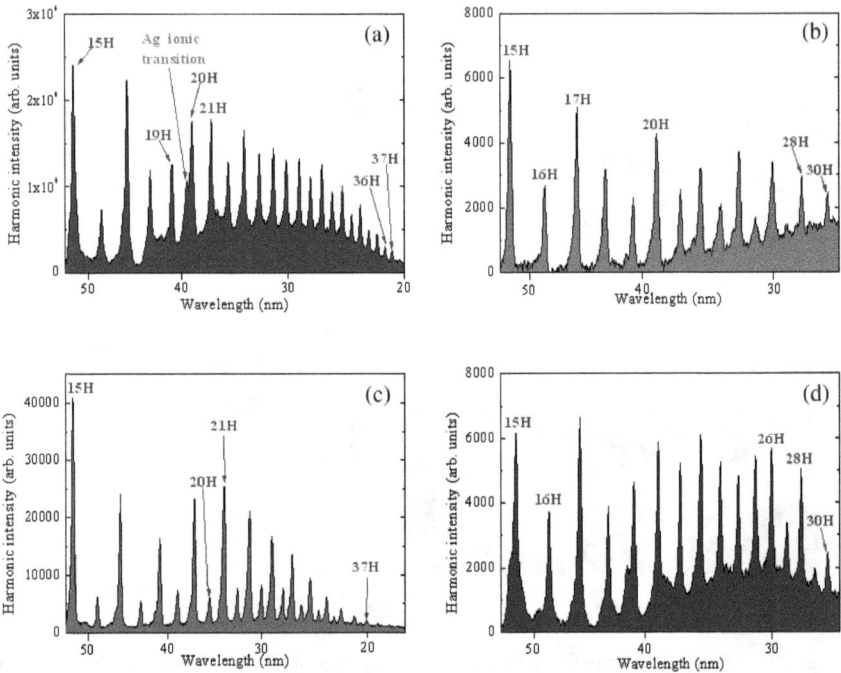

Fig. 3.8 (a) Harmonic spectra from silver plasma using the two-color pump configuration. (b) Optimization of even harmonics with regard to the odd ones in the cutoff region. (c) Harmonic spectra using a 200 mm focal length focusing mirror. (d) Harmonic spectra using a 500 mm focal length lens. Reprinted from [65] with permission from Optical Society of America.

The presence of all even harmonics is proof that sufficient overlap between the pulses occurs. Whilst the central parts of the two pump pulses were several pulse widths apart, there were still enough photons from the trailing edge of the second-harmonic pulse present in the plasma simultaneously with infrared photons to lead to appreciable wave-mixing effects. Otherwise there would only be harmonics every fourth order, e.g. 22nd, 26th, 30th and no 24th, 28th, etc. The influence of the focusing conditions for converting ω and 2ω waves on the distribution of odd and even harmonics was analyzed. The application of tight focusing (200 mm focal length mirror) led to generation of stronger odd harmonics, which were considerably distinguishable with regard to the even ones (Fig. 3.8c), while in the case of using the 500 mm focal length lens, one could generate equal odd and even harmonics (Fig. 3.8d).

An interesting feature of these spectra was the observation of an enhanced even (20th) harmonic (Fig. 3.8a), though its enhancement was not so pronounced compared to previously reported studies of single odd harmonics from some plasmas, probably due to the relatively moderate influence of the ionic resonance in the vicinity of the 20th harmonic on the nonlinear optical response.

The demonstrated peculiarities (extended tuning of the harmonics of broadband 780 nm radiation and resonance-induced growth of particular even harmonics) motivated a search for improvement of harmonic yield in two plasma plumes, chromium and vanadium, which have already shown advanced resonant-induced properties using low pulse repetition rate (10 Hz), narrowband (10 nm) laser sources [66, 67].

3.5.3. *Chromium plasma*

In the case of chromium plasma, two characteristic regimes of HHG were defined, which created different images of harmonic spectra. In the first case, at weak excitation of the target, a cutoff was observed at 31.2 nm ($E = 39.9$ eV, 25H). With an increase of target excitation and femtosecond pulse intensity inside the plume by moving the plasma towards the focus of 780 nm radiation, a second plateau appeared with characteristic strong 27th and 29th harmonics, and the cutoff was extended toward the range of $E \approx 59$ eV (37H). At chirp-free conditions, the 27th and 29th harmonics were approximately equal to each other. The variation of laser chirp, however, led to considerable change in the relative intensities of these two harmonics, while other harmonics remained approximately the same. Figure 3.9a shows the harmonic spectra in the range of 22–42 nm using different chirps of the driving radiation. The variation of chirp from +114 fs to −92 fs considerably improved the intensity of the 27th harmonic, while the 29th harmonic became weaker. This variation of harmonic intensities was attributed to the tuning of the 27th harmonic ($\lambda = 28.9$ nm, $E = 43.1$ eV at chirp-free conditions) towards the ionic resonance possessing strong oscillator strength. The presence of such a resonance was seen in the over-excitation of chromium plasma (see the ionic line at the right side of the 27th harmonic in Fig. 3.9b).

In the above-described studies (see Section 3.2), HHG in chromium plasma showed a considerable variation of the 27th harmonic intensity at

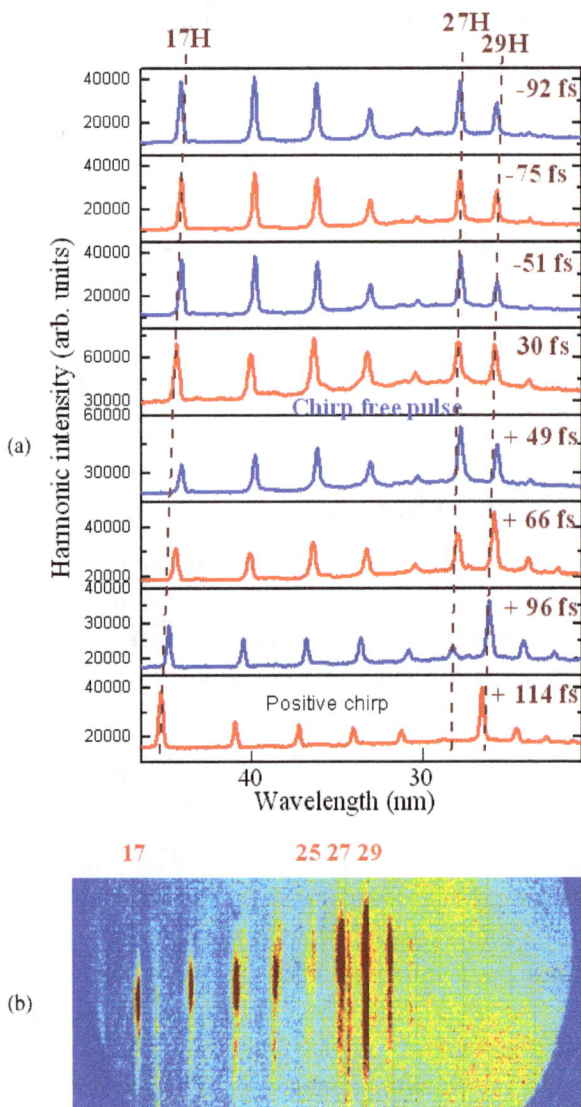

Fig. 3.9 (a) Harmonic spectra from chromium plasma at different chirps of laser radiation. Positive and negative values of pulse duration correspond to positively and negatively chirped pulses. (b) Harmonic spectrum at over-excited conditions of chromium plasma formation, with ionic lines appearing close to the enhanced 27th and 29th harmonics. Reprinted from [65] with permission from Optical Society of America.

different chirps of the driving radiation [66]. At some chirps, the 27th harmonic almost disappeared from the harmonic spectra. At the same time, a strong 29th harmonic was observed in the case of chirp-free pulses. Note that those studies were performed using 800 nm laser radiation. A small difference in laser wavelengths led in previous studies to closeness of the 29th harmonic and a strong resonance, which in the present research enhanced the 27th harmonic.

The calculations of gf in the photon energy range of 40–60 eV presented in [68] clearly show a group of transitions in the 44.5–44.8 eV region possessing very strong oscillator strengths (with gf varying between 1 and 2.2), which considerably exceeded those of other transitions in the range of 40–60 nm. These transitions are assumed to be responsible for the observed enhancement of the 27th and 29th harmonics. At the same time, strong photoabsorption lines within the 41–42 eV region reported in the above work may considerably decrease the yield of the 25th harmonic.

The application of a two-color pump also revealed the influence of the resonance ionic line on the efficiency of even harmonics generation. Their behavior was analogous to the odd ones. Even harmonics almost disappeared in the range of 24th to 26th harmonics (38.3–41.5 eV) in the case of chirp-free pulses, while the 28th harmonic (44.6 eV) was considerably stronger than the lowest-order harmonics and almost equal to the enhanced odd ones (Fig. 3.10a). The harmonic efficiency in the case of the two-color pump exceeded that observed in the case of the single-color pump (compare Figs. 3.10a and 3.10b).

3.5.4. Vanadium plasma

Finally, we show the results of studies of harmonic generation from vanadium plasma. The 71st harmonic of 800 nm radiation at photon energy of 110 eV was reported in [67], with a conversion efficiency of 1.6×10^{-7} by using a laser-ablation vanadium plume irradiated by a femtosecond laser pulse (see also Chapter 6). It was concluded, by measuring the spectra of the laser plasma and calculating the ionization conditions and harmonic cutoffs in the laser-ablation plume, that the higher harmonics originated from the interaction of the femtosecond laser pulses with doubly charged vanadium ions.

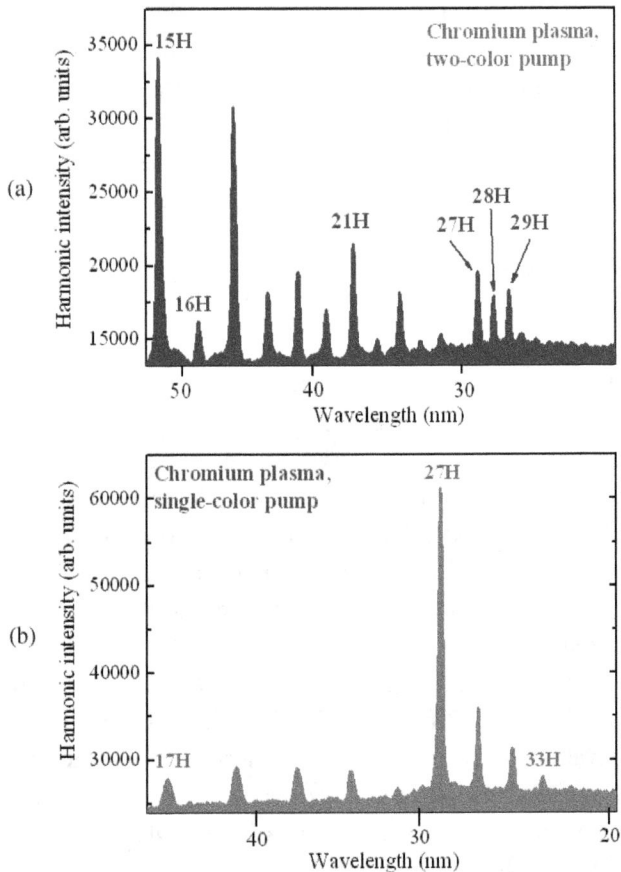

Fig. 3.10 (a) Two-color pump-induced spectra of harmonics from chromium plasma and (b) the spectra obtained at analogous experimental conditions by removing the second-harmonic crystal from the path of the 780 nm radiation. Reprinted from [65] with permission from Optical Society of America.

The dynamics of harmonic variations was the same as in the case of the chromium plasma. Initially, only low-order harmonics (up to the 23rd order) were obtained in the XUV spectra at a weak excitation of the vanadium target ($I_{pp} = 6 \times 10^9$ W cm^{-2}, Fig. 3.11a). With an increase of vanadium surface excitation ($I_{pp} \geq 1 \times 10^{10}$ W cm^{-2}), the spectrum showed the extension of harmonics, with the characteristic appearance of a "second plateau" beginning with the strong 27th harmonic (Fig. 3.11b). The two-color pump led to further improvement of harmonic yield from this plasma (Fig. 3.11c). It was followed

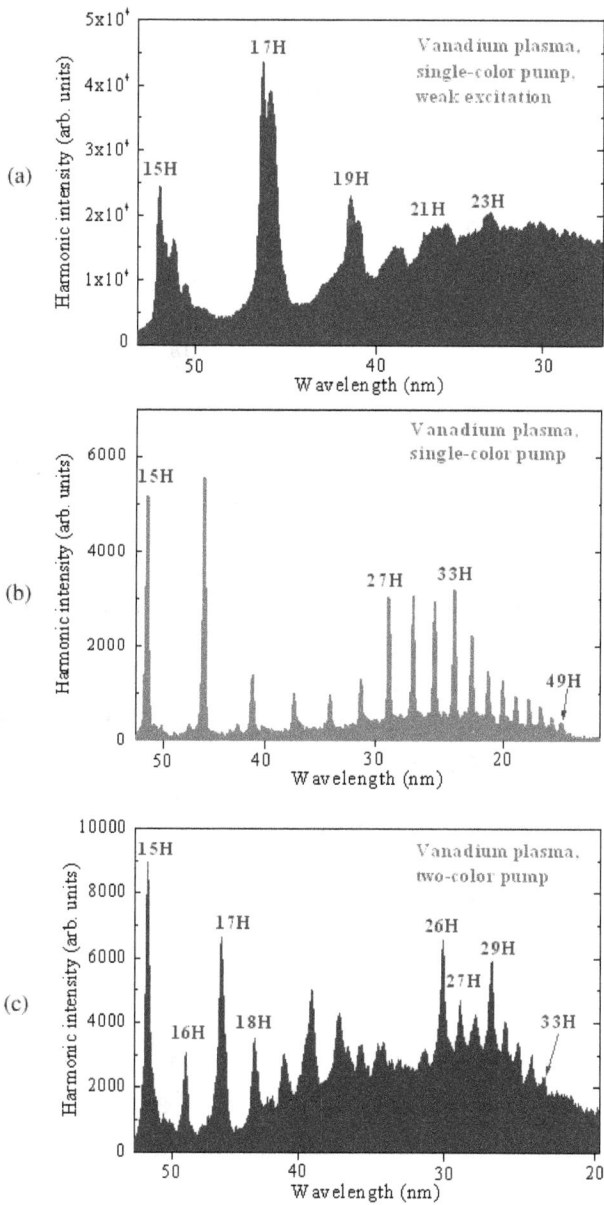

Fig. 3.11 Variations of harmonic spectra at (a) weak excitation of vanadium target ($I_{pp} = 6 \times 10^9 \, \mathrm{W \, cm^{-2}}$), (b) stronger excitation of target ($I_{pp} = 1 \times 10^{10} \, \mathrm{W \, cm^{-2}}$), and (c) application of two-color pump. Reprinted from [65] with permission from Optical Society of America.

by the appearance of an enhanced 26th harmonic ($\lambda = 30$ nm, $E = 41.5$ eV), which one can attribute to the influence of strong ionic transitions in this spectral range, analogous to the case of chromium plasma.

The important novelty of these studies is the application of a high pulse repetition source for resonance- and two-color-induced plasma HHG. This allowed the average power of generated XUV radiation to be improved by two orders of magnitude compared with the cases of 10 Hz laser sources used in previous plasma HHG studies for equal pulse energies. This improvement could be useful both for analysis of the temporal characteristics of generated harmonics and for the study of quantum path interference of contributing electron trajectories, both requiring long sets of measurements. More details on that topic are presented in Chapters 6 and 7.

During these studies, several new resonance-enhanced processes were revealed, mostly due to involvement of a weak second-harmonic wave, which modified the conditions of HHG in plasma plumes in such a manner that a change of electron trajectory at these conditions led to generation of both enhanced odd and even harmonics around different ranges of the plateau. Specifically, the observation of an enhanced 20th harmonic in silver plasma, 26th harmonic in vanadium plasma, and 28th harmonic in chromium plasma supported the mechanisms of the involvement of autoionizing states of atoms and ions in the process of enhancement of nonlinear optical processes in the vicinity of these resonances [44]. The important point here is the application of broadband radiation in short ($\sim$30 fs) pulses for plasma HHG, which allowed the fine tuning of harmonics along a broad range, thus giving an opportunity to control the optimal coincidence of the wavelength of autoionizing states and atomic and ionic resonances and some harmonics.

3.6. Theoretical Approaches for Description of Observed Peculiarities of Resonant Enhancement of Single Harmonic in Laser Plasma

The dependence of the recombination probability on the electron return energy and on the structure of the target is reflected in the HHG spectrum and has been the subject of intensive research in recent years. To enhance the notoriously low efficiency of the HHG process, it appears promising to exploit the effect of resonances, which are known to be of great importance in photoionization. The investigation of resonant peaks in the photoionization cross section has a long history, including studies of autoionization resonances [69, 70], shape resonances [71], and giant resonances [72], but there have been only a few studies on the role of resonances in HHG. The role of atomic resonances in increasing the laser radiation conversion efficiency was actively discussed in the framework of perturbation theory at the early stages of the study of low-order harmonic generation (see monograph [73] and the references therein). In the case of HHG, the increase in efficiency of generated harmonics due to resonance processes came under discussion a decade ago, and this approach appears to have considerable promise with the use of ionic and, in some cases, atomic resonances [10]. Those papers comprise both a theoretical treatment of the process and a description of the first attempts to form resonance conditions in experiments. While theoretical estimates attested the possibility of efficient enhancement of individual harmonics and groups of harmonics, experimental works revealed the difficulties encountered in HHG in gases. Therefore, the use of plasma media could largely facilitate the solution to the problem of resonance harmonic enhancement. Examining a large group of potential targets allowed identification of some of them as suitable for demonstrating this process [18, 51]. The advantages of "plasma HHG" over "gas HHG" were amply manifested in this case, because the number of possible media in the former case is far greater than in the latter case.

As already mentioned, attempts to explain experimental observations of resonant enhancement in plasma harmonics have been reported recently in [43–50, 74–76]. In particular, in [43], it has been shown that the influence of atomic autoionizing states on the phase matching of HHG may result in efficient selection of the single harmonic in calcium plasma. This was the first report of efficient high-order harmonic selection using autoionizing states. The calculations [43] show that the achievement of phase matching for HHG of Ti:sapphire laser radiation in Ca^+ plasma results in the selection of a single (21st) harmonic with a contrast of 10^4 and conversion efficiency of $\sim 10^{-3}$. The contrast of harmonic selection depends on the medium density and bandwidth of the laser radiation. The variations of the plasma components and fundamental wavelength result in the tuning of a selected harmonic frequency in the plateau region. The influence of the AC Stark shift and free electrons changes the phase mismatch and the optimal laser frequency at which the efficient selection of a single harmonic is achieved. So, the real intensity enhancement due to propagation effects can be even greater than that in the single-atom approximation.

An approach that suggests an HHG model describing enhancement of the generation efficiency for the harmonic resonant with the transition between the ground and the autoionizing state of the generating ion was developed by Strelkov [44]. In his model, the third (recombination) step of the three-step scenario is partitioned into two steps: the capture of a laser-accelerated electron into an autoionizing state of the parent ion followed by the radiative relaxation of this state to the ground state with emission of the harmonic photon.

Figure 3.12a shows calculations from [44] indicating that while the enhancement values for different media differ by almost two orders of magnitude, the theoretical results are close to the experimental ones. The difference between them is attributable to the medium effects (harmonic absorption and detuning from HHG phase matching) that are not taken into account in this theory.

Although this four-step model provides reasonable estimates for the ratio of the enhanced harmonic intensity to the averaged intensity of neighboring harmonics, the authors of [47] point out that the suppression of harmonics preceding the resonant one remains a puzzle for the theoretical model [44]. In their research [47], they show that enhancements of single harmonics with harmonic energies near the energies of autoionizing states in atoms or atomic ions, as well as the aforementioned suppression of the preceding (or, in some cases, the succeeding) harmonics, may be interpreted (at least for those harmonics in the region of the classical plateau cutoffs for a given laser frequency and intensity) in terms of the usual three-step scenario for the HHG process [77, 78], without any additional assumptions or extensions. In particular, they successfully reproduced the main features observed in experiments on HHG from plasmas produced by the laser ablation of solid chromium and manganese targets. For both 800 and 400 nm wavelengths, these features are caused by atomic structure effects in the radiative recombination cross sections of Cr^{2+} and Mn^{2+} ions (or, equivalently, in the photoionization cross sections of Cr^+ and Mn^+ ions). These effects were predicted by the factorization formula [79] for HHG rates. The experimental measurements of these rates serve to

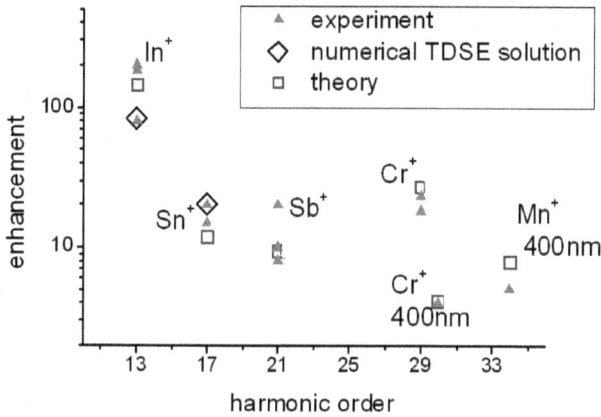

(a)

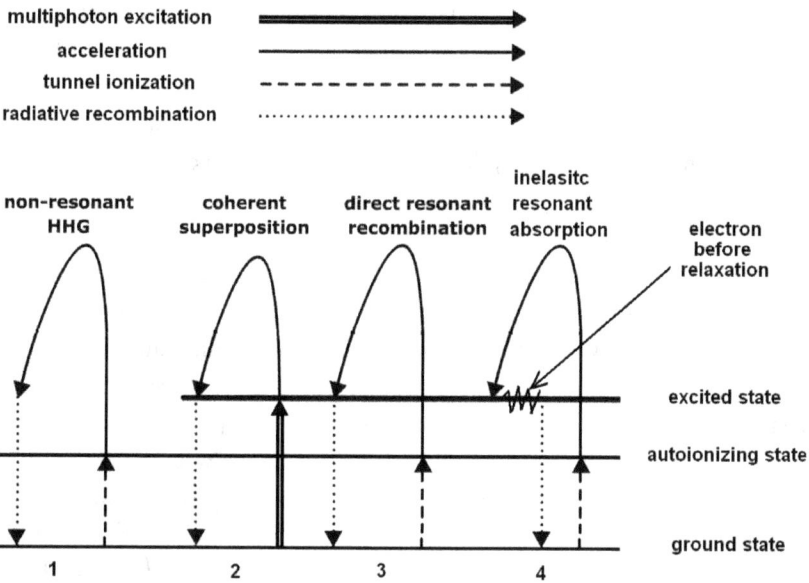

(b)

Fig. 3.12 (a) Comparison of the experimental results on resonance enhancement of harmonics and analytical and numerical results. Adapted from [44] with permission from American Physical Society. (b) Various scenarios of resonant and nonresonant HHG: 1, nonresonant HHG (ordinary three-step process); 2, HHG through resonant multiphoton excitation [45, 46]; 3, HHG from direct resonant recombination [50]; 4, HHG from relaxation after inelastic absorption of an electron by an excited state [44]. Reprinted from [50] with permission from Optical Society of America.

complement the measurements of Cooper minima in HHG from neutral atoms [80, 81].

It was found in [46] that the laser intensity dependence of the intensity and phase of the single harmonic generated in resonant HHG from plasma ablation is different from that of the standard plateau and cutoff high harmonics. The resonant harmonic intensity increases continuously (i.e., without rapid oscillations) with the increase of the laser intensity, while the resonant harmonic phase is almost constant. For HHG, such unusual behavior of the harmonic phase requires a detailed experimental investigation. Namely, the harmonic phase dependence is important for synchronization of high-order harmonics. The subfemtosecond light pulses can be obtained by superposing several high-order harmonics.

In the context of a recent first attosecond pulse train reconstruction of high-order harmonics from laser ablation plasma [82] the results of [46] are even more important. It was found that the temporal profile of a group of odd harmonics, which encompasses the resonant harmonic, is in the form of a broad peak in each laser-field half-cycle. This is an advantage in comparison with the usual plateau and cutoff harmonics where two such peaks are generated per half-cycle, which requires an appropriate experimental technique (i.e., focusing such that the collective effects due to macroscopic propagation select only one peak). Taking into account a smooth dependence of the harmonic intensity on the laser intensity and that it is not necessary to manipulate with long and short orbits by appropriate focusing, one can expect that resonant HHG has bright prospects for application in attoscience [83, 84].

The HHG in the presence of a shape resonance was analyzed in [49]. To understand the HHG mechanism, the time–frequency analysis of the intensity and phase was investigated. It was found that the resonance gives rise to a clear signature in the HHG spectrum irrespective of the pulse length. The time–frequency analysis supports Strelkov's four-step model. While the present one-dimensional calculation favors capture from the long trajectory, one can expect that a full three-dimensional calculation will show a similar mechanism, but with higher weight given to the short trajectory. By the nature of this process, the emitted harmonic radiation is phase locked with the usual harmonic emission from the short and long trajectories. For long-lived resonances, interference occurs between the populations caused by recollisions in different half-cycles. This leads to new possibilities for XUV pulse shaping on the subfemtosecond timescale.

In [48], simulations of resonant HHG were performed by means of a multiconfigurational time-dependent Hartree–Fock (MCTDHF) approach for three-dimensional fullerene-like systems and the influence of the surface plasmon resonance (SPR) of C_{60} on harmonic efficiency in the range of 60 nm ($E = 20\,eV$) was analyzed. These results showed the methods of resonant HHG optimization and, most importantly, attosecond pulse train generation. The MCTDHF calculations of HHG from C_{60} clusters were in good qualitative agreement with experimental data reported in previous studies of harmonic generation in fullerene-containing laser plumes. The broadness of the SPR in C_{60} allows a direct stimulated transition from the continuum into the ground state without additional need of a radiationless transition, thus making possible competing enhancement of neighboring harmonics, which is useful for attosecond pulse train generation. In general, multi-electron SPR of C_{60} is a generalization to two-electron autoionizing states in atoms and simple molecules. However, the extreme width of the plasmon resonance allows direct recombination, whereas for autoionizing states radiationless transition to these states should happen first. Use of a strongly ionized medium with some delocalized electrons as a target for resonant HHG can be favorable for extension of such attosecond pulse trains into the water-window region.

The work [50] was devoted to finding an explanation of the observed phenomenon of resonant enhancement of a single harmonic in indium plasma without enhancement of neighboring ones. It can also be used to predict the most promising targets for resonant HHG and to increase its efficiency by control of the pump radiation parameters. As the driving electric field amplitudes needed for this process are close to intra-atomic ones and the pump laser field can no longer be treated as a small perturbation, resonant HHG becomes nonperturbative, and the time-independent methods, such as perturbation theory, are no longer applicable. A promising way to investigate this process is to solve numerically a time-dependent Schrödinger equation in some approximation for a given system. This means that the wave functions of the system and their evolution are computed only at certain points in real space, called the grid points, and they are extrapolated to other regions of space. All the computations in this work [50] were performed using time-dependent density functional theory (TDDFT) [85] with the aid of real-space real-time code OCTOPUS [86, 87], which is a software package for performing Kohn–Sham TDDFT calculations. A detailed description of the TDDFT formalism can be found in [88]. The results of TDDFT calculations of HHG for the indium ion were found to be in good qualitative correlation with experimental data [14]. This allowed proof of the possibility of direct resonant recombination for HHG on the basis of calculations for artificially chirped pulses. The peculiarities of this approach were discussed and compared with existing theories of resonant enhancement of harmonics.

In Fig. 3.12b, some of the aforementioned theoretical approaches for explanation of observed resonance enhancement of harmonics [44, 46, 50] are schematically compared. Figure 3.12b, scenario 1 shows the nonresonant three-step process, when there are no resonant states in the system; it is shown for comparison. A key feature of [46] (Fig. 3.12b, scenario 2) is the population of the resonant excited states via multiphoton ionization. It should be pointed out that multiphoton absorption is naturally incorporated into TDDFT simulations. However, the numerical simulations [50] show that resonant HHG is observed when resonant conditions are fulfilled for the intensities that bring the system into the tunnel ionization regime. This means that, at least in these simulations, the approaches that rely on resonant multiphoton absorption as the main ionization mechanism are not reliable.

In theory [45, 46], resonant HHG originates from electrons starting from the excited state that was populated by a multiphoton resonance and recombining into the ground state. One can easily see that this process can lead to simultaneous resonant enhancement of several neighboring harmonics.

On the other hand, in experiments [89] the harmonics closest to a resonant one are suppressed even compared to the nonresonant case. It can easily be explained by means of resonant recombination theory (RRT) that the neighboring harmonics should be decreased. Indeed, when an electron starts from the ground state and recombines into it again, the increase in the probability of recombination with emission of a certain harmonic reduces the probability of other recombinations. So, the main advantage of the theory of resonant recombination lies in the fact that it allows explanation of resonant enhancement of a single harmonic.

Another fascinating peculiarity of RRT is the fact that all the equations given in [45, 46] can also describe RRT theory without any changes. The only difference lies in the fact that the additional resonant term

$$\sum_{j,j'} a_j^* a_{j'} \langle j|r|j' \rangle \exp[-i(E_{j'} - E_j)t] \qquad (3.1)$$

in RRT not merely increases the effectiveness of ionization, as considered in [45, 46], but plays the decisive role in the resonant recombination process. So this theory can easily be explained semiclassically as well, giving full effectiveness of the semiclassical approach.

At the same time, resonant recombination is not a simple lasing process, as nonresonant nonperturbative HHG lies at its origin. In addition, resonant recombination does not require an exact multiphoton resonance. Facts prove that resonant HHG is most efficient in the case of the tunneling, not the multiphoton, ionization regime. As an important consequence of RRT, the most promising targets for resonant HHG should have high-lying excited states, which have a compromise between spontaneous radiation probability and resonance lifetime. This is fulfilled best of all in systems having delocalized electrons, such as atoms with weakly screened d-electrons, metal nanoparticles, and molecules of fullerenes.

Another important consequence of RRT is the necessity to get the best resonant conditions in the region of the pulse having sufficiently high intensity, that is, in the tunnel ionization regime. At the same time, theory [45,46] would require the best resonant conditions in the low-intensity regions of the pulse, because only at relatively low intensities is multiphoton excitation most probable. Such a difference allows us to check the correctness of RRT and theory [45, 46] both theoretically and experimentally by investigating the spectral distribution of energy within the pulse.

As multiphoton absorption is possible only at relatively low intensities, while tunnel ionization takes place at higher intensities, multiphoton processes are unlikely to play a major role in resonant recombination. At the same time, resonant recombination needs exact resonance mainly at higher intensities, so RRT simulations may be its first confirmation. The optimization of resonant HHG is thus influenced most by the parts of the pump pulse with the highest intensity and needs better tuning to resonance than simple compression of the whole pulse, which has little spectral selectivity.

Resonant recombination theory computations have been found to be in good qualitative agreement with the experimental results of fullerene HHG [90]. This allowed the theory of resonant recombination to be applied for the system where the probability of an electron's recombination into the ground state is greatly multiplied when the sum of its kinetic energy acquired from the laser field and its ionization energy equals the energy difference between the resonant and ground-state levels.

It is much more difficult to distinguish the RRT approach [50] (Fig. 3.12b, scenario 3) from [44] (Fig. 3.12b, scenario 4). An important physical difference between these approaches is the population of the excited state. In [44], it should be sufficiently large for a mean spontaneous recombination (relaxation) time. Otherwise, it would be a simple lasing, and the conversion efficiency could grow with the increase of the length of the nonlinear media, although this is not proved by experiments [14]. On the other hand, direct resonant recombination does not impose this restriction on state population. Population of states can in principle be visualized, but such a possibility is in the test stage, and there is also no population monitoring in [44]. An observation of resonant HHG itself in any time-dependent run cannot distinguish the corresponding approaches in their current formulation. It is also possible that both approaches are correct but under different conditions. Most conclusions derived in [44] are also valid for the [50] approach.

As a general conclusion, in any time-dependent HHG calculation that supports strong excited states, a resonant HHG should be observed if the resonant conditions are met at the moment of recombination. The states themselves can be artificially introduced for single-electron models [44] or can follow naturally from the potential-well structure in multi-electron calculations. The scheme of direct resonant recombination presented in Fig. 3.12b for the corresponding initial system and +1 ionized initial system should explain all such phenomena. As a result, the method used to solve the approximated Schrödinger equation can have only a quantitative effect on the system. The article [91], completed within the TDDFT framework, is clear evidence of this idea.

It should be mentioned that direct application of the MCTDHF method to large systems is still too demanding because of the exponential growth of computational resources required with the increase of the number of particles. According to the qualitative independence of resonant HHG observations from the method of solution of the approximated Schrödinger equation for a given system, the approaches, such as TDDFT, that scale almost linearly with the number of particles are quite valid to be chosen for further investigation of complicated multielectron systems. It is worth noting that recent experimental observations of resonance enhancement of harmonics [92] point to the importance of consideration of various mechanisms of harmonic enhancement.

References

1. Chang, Z., Rundquist, A., Wang, H. *et al.* (1997). Generation of coherent soft X-rays at 2.7 nm using high harmonics, *Phys. Rev. Lett.*, 79, 2967–2970.
2. Gisselbrecht, M., Descamps, D., Lynga, C. *et al.* (1999). Absolute photoionization cross sections of excited He states in the near-threshold region, *Phys. Rev. Lett.*, 82, 4607–4610.
3. Bauer, M., Lei, C., Read, K. *et al.* (2001). Direct observation of surface chemistry using ultrafast soft-X-ray pulses, *Phys. Rev. Lett.*, 87, 025501.
4. Nagasono, M., Suljoti, E., Pietzsch, A. *et al.* (2007). Resonant two-photon absorption of extreme-ultraviolet free-electron-laser radiation in helium, *Phys. Rev. A*, 75, 051406.
5. Paul, P.M., Toma, E.S., Breger, P. *et al.* (2001). Observation of a train of attosecond pulses from high harmonic generation, *Science*, 292, 1689–1692.
6. L'Huillier, A., Descamps, D., Johansson, A. *et al.* (2003). Applications of high-order harmonics, *Eur. Phys. J. D*, 26, 91–98.
7. Rundquist, A., Durfee, C.G., Chang, Z.H. *et al.* (1998). Phase-matched generation of coherent soft X-rays, *Science*, 280, 1412–1415.
8. Kazamias, S., Douillet, D., Weihe, F. *et al.* (2003). Global optimization of high harmonic generation, *Phys. Rev. Lett.*, 90, 193901.
9. Figueira de Morisson Faria, C., Copold, R., Becker, W. *et al.* (2002). Resonant enhancements of high-order harmonic generation, *Phys. Rev. A*, 65, 023404.
10. Taieb, R., Veniard, V., Wassaf, J. *et al.* (2003). Roles of resonances and recollisions in strong-field atomic phenomena. II. High-order harmonic generation, *Phys. Rev. A*, 68, 033403.
11. Gibson, E.A., Paul, A., Wagner, N. *et al.* (2003). Coherent soft X-ray generation in the water window with quasi-phase matching, *Science*, 302, 95–98.
12. Suzuki, M., Baba, M., Ganeev, R. *et al.* (2006). Anomalous enhancement of single high-order harmonic using laser ablation tin plume at 47 nm, *Opt. Lett.*, 31, 3306–3308.
13. Suzuki, M., Baba, M., Kuroda, H. *et al.* (2007). Intense exact resonance enhancement of single-high-harmonic from an antimony ion by using Ti:sapphire laser at 37 nm, *Opt. Express*, 15, 1161–1166.
14. Ganeev, R.A., Suzuki, M., Ozaki, T. *et al.* (2006). Strong resonance enhancement of a single harmonic generated in extreme ultraviolet range, *Opt. Lett.*, 31, 1699–1701.
15. Chang, Z., Rundquist, A., Wang, H. *et al.* (1998). Temporal phase control of soft-x-ray harmonic emission, *Phys. Rev. A*, 58, R30–R33.

16. Kim, H.T., Lee, D.G., Hong, K.H. *et al.* (2003). Continuously tuneable high-order harmonics from atoms in an intense femtosecond laser field, *Phys. Rev. A*, **67**, 051801.

17. Kim, H.T., Kim, J.-H., Lee, D.G. *et al.* (2004). Optimization of high-order harmonic brightness in the space and time domains, *Phys. Rev. A*, **69**, 031805.

18. Ganeev, R.A., Naik, P.A., Singhal, H. *et al.* (2007). Strong enhancement and extinction of single harmonic intensity in the mid- and end-plateau regions of the high harmonics generated in low-excited laser plasmas, *Opt. Lett.*, **32**, 65–67.

19. Ganeev, R.A., Suzuki, M., Redkin, P.V. *et al.* (2007). Variable pattern of high harmonic spectra from a laser-produced plasma by using the chirped pulses of narrow-bandwidth radiation, *Phys. Rev. A*, **76**, 023832.

20. Suzuki, M., Baba, M., Ganeev, R.A. *et al.* (2007). Observation of single harmonic enhancement due to quasi-resonance conditions with the tellurium ion transition in the range of 29.44 nm, *J. Opt. Soc. Am. B*, **24**, 2686–2689.

21. Ganeev, R.A., Singhal, H., Naik, P.A. *et al.* (2009). Variation of harmonic spectra in laser-produced plasmas at variable phase modulation of femtosecond laser pulses of different bandwidth, *J. Opt. Soc. Am. B*, **26**, 2143–2151.

22. Ganeev, R.A., Suzuki, M., Baba, M. *et al.* (2005). Generation of strong coherent extreme ultraviolet radiation from the laser plasma produced on the surface of solid targets, *Appl. Phys. B*, **81**, 1081–1089.

23. Ganeev, R.A., Singhal, H., Naik, P.A. *et al.* (2006). Harmonic generation from indium-rich plasmas, *Phys. Rev. A*, **74**, 063824.

24. Ganeev, R.A., Elouga Bom, L.B., Kieffer, J.-C. *et al.* (2007). Optimum plasma conditions for the efficient high-order harmonic generation in platinum plasma, *J. Opt. Soc. Am. B*, **24**, 1319–1323.

25. Ganeev, R.A., Elouga Bom, L.B., Kieffer, J.-C. *et al.* (2007). Systematic investigation of resonance-induced single harmonic enhancement in the extreme ultraviolet range, *Phys. Rev. A*, **75**, 063806.

26. Duffy, G. and Dunne, P. (2001). The photoabsorption spectrum of an indium laser produced plasma, *J. Phys. B: At. Mol. Opt. Phys.*, **34**, L173–L178.

27. Ganeev, R.A., Naik, P.A., Singhal, H. *et al.* (2007). Tuning of the high-order harmonics generated from laser plasma plumes and solid surfaces by varying the laser spectrum, chirp, and focal position, *J. Opt. Soc. Am. B*, **24**, 1138–1143.

28. Ganeev, R.A., Suzuki, M., Baba, M. *et al.* (2005). Harmonic generation in XUV from chromium plasma, *Appl. Phys. Lett.*, **86**, 131116.

29. Ganeev, R.A., Singhal, H., Naik, P.A. *et al.* (2006). Single harmonic enhancement by controlling the chirp of the driving laser pulse during high-order harmonic generation from GaAs plasma, *J. Opt. Soc. Am. B*, **23**, 2535–2540.

30. Milošević, D.B. and Becker, W. (2002). Role of long quantum orbits in high-order harmonic generation, *Phys. Rev. A*, **66**, 063417.

31. Kubodera, S., Nagata, Y., Akiyama, Y. *et al.* (1993). High-order harmonic generation in laser-produced ions, *Phys. Rev. A*, **48**, 4576–4582.

32. Ganeev, R.A., Elouga Bom, L.B., Kieffer, J.-C. *et al.* (2007). Demonstration of the 101st harmonic generated from laser-produced manganese plasma, *Phys. Rev. A*, **76**, 023831.

33. Shan, B., Cavalieri, A. and Chang, Z. (2002). Tuneable high harmonic generation with an optical parametric amplifier, *Appl. Phys. B*, **74**, S23–S26.

34. Tosa, V., Kim, H.T., Kim, I.J. *et al.* (2005). High-order harmonic generation by chirped and self-guided femtosecond laser pulses. II. Time-frequency analysis, *Phys. Rev. A*, 71, 063808.
35. Froud, C.A., Rogers, E.T.F., Hanna, D.C. *et al.* (2006). Soft-x-ray wavelength shift induced by ionization effects in a capillary, *Opt. Lett.*, 31, 374–376.
36. Reitze, D.H., Kazamias, S., Weihe, F. *et al.* (2004). Enhancement of high-order harmonic generation at tuned wavelengths through adaptive control, *Opt. Lett.*, 29, 86–88.
37. Kim, K.Y., Alexeev, I., Antonsen, T.M. *et al.* (2005). Spectral redshifts in the intense laser-cluster interaction, *Phys. Rev. A*, 71, 011201.
38. Toma, E.S., Antoine, P., de Bohan, A. *et al.* (1999). Resonance-enhanced high-harmonic generation, *J. Phys. B: At. Mol. Opt. Phys.*, 32, 5843–5852.
39. Xu, H., Tang, X. and Lambropoulos, P. (1992). Nonperturbative theory of harmonic generation in helium under a high-intensity laser field: The role of intermediate resonance and of the ion, *Phys. Rev. A*, 46, 2225–2228.
40. Gaarde, M.B. and Schafer, K.J. (2001). Enhancement of many high-order harmonics via a single multiphoton resonance, *Phys. Rev. A*, 64, 013820.
41. Zeng, Z., Li, R., Cheng, Y. *et al.* (2002). Resonance-enhanced high-order harmonic generation and frequency mixing in two-colour laser field, *Phys. Scripta*, 66, 321–325.
42. Bartels, R., Backus, S., Zeek, E. *et al.* (2000). Shaped-pulse optimization of coherent emission of high-harmonic soft X-rays, *Nature*, 406, 164–166.
43. Kulagin, I.A. and Usmanov, T. (2009). Efficient selection of single high-order harmonic caused by atomic autoionizing state influence, *Opt. Lett.*, 34, 2616–2619.
44. Strelkov, V. (2010). Role of autoionizing state in resonant high-order harmonic generation and attosecond pulse production, *Phys. Rev. Lett.*, 104, 123901.
45. Milošević, D.B. (2007). High-energy stimulated emission from plasma ablation pumped by resonant high-order harmonic generation, *J. Phys. B: At. Mol. Opt. Phys.*, 40, 3367–3376.
46. Milošević, D.B. (2010). Resonant high-order harmonic generation from plasma ablation: Laser intensity dependence of the harmonic intensity and phase, *Phys. Rev. A*, 81, 023802.
47. Frolov, M.V., Manakov, N.L. and Starace, A.F. (2010). Potential barrier effects in high-order harmonic generation by transition-metal ions, *Phys. Rev. A*, 82, 023424.
48. Redkin, P.V. and Ganeev, R.A. (2010). Simulation of resonant high-order harmonic generation in three-dimensional fullerenelike system by means of multiconfigurational time-dependent Hartree-Fock approach, *Phys. Rev. A*, 81, 063825.
49. Tudorovskaya, M. and Lein, M. (2011). High-order harmonic generation in the presence of a resonance, *Phys. Rev. A*, 84, 013430.
50. Redkin, P.V., Kodirov, M.K. and Ganeev, R.A. (2011). Investigation of resonant nonperturbative high-order harmonic generation in indium vapors, *J. Opt. Soc. Am. B*, 28, 165–170.
51. Ganeev, R.A. (2009). Application of resonance-induced processes for enhancement of the high-order harmonic generation in plasma, *Open Spectrosc. J.*, 3, 1–8.
52. Ganeev, R.A. (2007). High-order harmonic generation in laser plasma: A review of recent achievements. *J. Phys. B: At. Mol. Opt. Phys.*, 40, R213–R253.
53. Ganeev, R.A. (2009). Generation of high-order harmonics of high-power lasers in plasmas produced under irradiation of solid target surfaces by a prepulse, *Phys. Usp.*, 52, 55–77.

54. Chen, G., Chen, J.G., Yang, Y.J. *et al.* (2010). Effect of the relative phase between two-colour pump pulses on structure of harmonic spectra, *Eur. Phys. J. D*, 57, 145–149.

55. Cormier, E. and Lewenstein, M. (2000). Optimizing the efficiency in high order harmonic generation optimization by two-colour fields, *Eur. Phys. J. D*, 12, 227–233.

56. Kim, I.J., Kim, C.M., Kim, H.T. *et al.* (2005). Highly efficient high-harmonic generation in an orthogonally polarized two-color laser field, *Phys. Rev. Lett.*, 94, 243901.

57. Mauritsson, J., Johnsson, P., Gustafsson, E. *et al.* (2006). Attosecond pulse trains generated using two color laser fields, *Phys. Rev. Lett.*, 97, 013001.

58. Pfeifer, T., Gallmann, L. and Abel, M.J. (2006). Single attosecond pulse generation in the multicycle-driver regime by adding a weak second-harmonic field, *Opt. Lett.*, 31, 975–977.

59. Charalambidis, D., Tzallas, P., Benis, E.P. *et al.* (2008). Exploring intense attosecond pulses, *New J. Phys.*, 10, 025018.

60. Ganeev, R.A., Singhal, H., Naik, P.A. *et al.* (2009). Enhancement of high-order harmonic generation using two-color pump in plasma plumes, *Phys. Rev. A*, 80, 033845.

61. Ganeev, R.A., Singhal, H., Naik, P.A. *et al.* (2010). Systematic studies of two-color pump induced high order harmonic generation in plasma plumes, *Phys. Rev. A*, 82, 053831.

62. Ganeev, R.A., Chakera, J.A., Naik, P.A. *et al.* (2011). Resonance enhancement of single even harmonic in tin-containing plasma using intensity variation of two-color pump, *J. Opt. Soc. Am. B*, 28, 1055–1061.

63. Pont, M. (1989). Atomic distortion and ac-Stark shifts of H under extreme radiation conditions, *Phys. Rev. A*, 40, 5659.

64. Ganeev, R.A., Hutchison, C., Siegel, T. *et al.* (2011). High-order harmonic generation from metal plasmas using 1 kHz laser pulses, *J. Modern Opt.*, 58, 819–824.

65. Ganeev, R.A., Hutchison, C., Zaïr, A. *et al.* (2012). Enhancement of high harmonics from plasmas using two-color pump and chirp variation of 1 kHz Ti:sapphire laser pulses, *Opt. Express.*, 20, 90–100.

66. Ganeev, R.A. (2012). Harmonic generation in laser-produced plasma containing atoms, ions and clusters: A review, *J. Modern Opt.*, 59, 409–439.

67. Suzuki, M., Baba, M., Kuroda, H. *et al.* (2007). Seventy first harmonic generation from ions in laser ablation vanadium plume at 11.2 nm, *Opt. Express*, 15, 4112–4117.

68. McGuinness, C., Martins, M., Wernet, P. *et al.* (1999). Metastable state contributions to the measured 3p photoabsorption spectrum of Cr^+ ions in a laser-produced plasma, *J. Phys. B: At. Mol. Opt. Phys.*, 32, L583–L592.

69. Fano, U. (1961). Effects of configuration interaction on intensities and phase shifts, *Phys. Rev.*, 124, 1866–1882.

70. Raşeev, G., Leyh, B. and Lefebvre-Brion, H. (1986). Autoionization in diatomic molecules: An example of electrostatic autoionization in CO, *Z. Phys. D*, 2, 319–326.

71. Keller, F. and Lefebvre-Brion, H. (1986). Shape resonances in photoionization of diatomic molecules: An example in the d inner shell ionization of the hydrogen halides, *Z. Phys. D*, 4, 15 (1986).

72. Amusia, M.Y. and Connerade, J.-P. (2000). The theory of collective motion probed by light, *Rep. Prog. Phys.*, 63, 41–70.

73. Reintjes, J.F. (1984). *Nonlinear Optical Parametric Processes in Liquids and Gases*, Academic Press, New York.

74. Ganeev, R.A. and Redkin, P.V. (2008). Role of resonances in the high-order harmonic enhancement in diatomic molecules, *Opt. Commun.*, **281**, 4126–4129.
75. Ganeev, R.A. and Milošević, D.B. (2008). Comparative analysis of the high-order harmonic generation in the laser ablation plasmas prepared on the surfaces of complex and atomic targets, *J. Opt. Soc. Am. B*, **25**, 1127–1134.
76. Redkin, P.V., Danailov, M. and Ganeev, P.A. (2011). Endohedral fullerenes: A way to control resonant high-order harmonic generation, *Phys. Rev. A*, **84**, 013407.
77. Schafer, K.J., Yang, B., DiMauro, L.F. *et al.* (1993). Above threshold ionization beyond the high harmonic cutoff, *Phys. Rev. Lett.*, **70**, 1599–1602.
78. Corkum, P.B. (1993). Plasma perspective on strong field multiphoton ionization, *Phys. Rev. Lett.*, **71**, 1994–1997.
79. Frolov, M.V., Manakov, N.L., Sarantseva, T.S. *et al.* (2009). Analytic description of the high-energy plateau in harmonic generation by atoms: Can the harmonic power increase with increasing laser wavelengths? *Phys. Rev. Lett.*, **102**, 243901.
80. Minemoto, S., Umegaki, T., Oguchi, T. *et al.* (2008). Retrieving photorecombination cross sections of atoms from high-order harmonic spectra, *Phys. Rev. A*, **78**, 061402.
81. Wörner, H.J., Niikura, H., Bertrand, J.B. *et al.* (2009). Observation of electronic structure minima in high-harmonic generation, *Phys. Rev. Lett.*, **102**, 103901.
82. Elouga Bom, L.B., Haessler, S., Gobert, O. *et al.* (2011). Attosecond emission from chromium plasma, *Opt. Express*, **19**, 3677–3685.
83. Agostini, P. and DiMauro, L.F. (2004). The physics of attosecond light pulses, *Rep. Prog. Phys.*, **67**, 813–856.
84. Krausz, F. and Ivanov, M. (2009). Attosecond physics, *Rev. Mod. Phys.*, **81**, 163–234.
85. Runge, E. and Gross, E.K.U. (1984). Density-functional theory for time-dependent systems, *Phys. Rev. Lett.*, **52**, 997–1000.
86. Marques, M.A.L., Castro, A., Bertsch, G.F. *et al.* (2003). Octopus: a first-principles tool for excited electron–ion dynamics, *Comput. Phys. Commun.*, **151**, 60–78.
87. Castro, A., Appel, H., Oliveira, M. *et al.* (2006). Octopus: A tool for the application of time-dependent density functional theory, *Phys. Stat. Sol. B*, **243**, 2465–2488.
88. Marques, M.A.L., Ullrich, C.A., Nogueira, F., *et al.* (2003). *Time-Dependent Density Functional Theory*, Springer-Verlag, Berlin–Heidelberg.
89. Ganeev, R.A. (2012). Generation of harmonics of laser radiation in plasmas, *Laser Phys. Lett.*, **9**, 175–194.
90. Ganeev, R.A., Elouga Bom, L.B., Wong, M.C.H. *et al.* (2009). High-order harmonic generation from C_{60}-rich plasma, *Phys. Rev. A*, **80**, 043808.
91. Ruggenthaler, M., Popruzhenko, S.V. and Bauer, D. (2008). Recollision-induced plasmon excitation in strong laser fields, *Phys. Rev. A*, **78**, 033413.
92. Ganeev, R.A., Strelkov, V.V., Hutchison, C. *et al.* (2012). Experimental and theoretical studies of two-color pump resonance-induced enhancement of odd and even harmonics from a tin plasma, *Phys. Rev. A*, **85**, 023832.

4

Cluster-Containing Plasma Plumes: Attractive Media for High-Order Harmonic Generation of Laser Radiation

The nonlinear optical properties of nanoparticles have attracted much attention due to their potential applications in optoelectronics, mode-locking technologies, optical limiting, etc. Currently, nanoparticle formation during laser ablation of solid-state targets using ultrashort (femtosecond) laser pulses is a widely accepted technique. Together with the formation of ripples of wavelength range sizes, this technique provides the opportunity to create exotic structures with variable physical and chemical properties. The increase of surface area of the nanostructured materials allows for enhancement of the speed of catalytic reactions, and provides opportunities for application of such structures for information writing, manufacturing of lubricants, semiconductor technologies, etc. The structural, optical, and nonlinear optical parameters of nanoparticles are known to differ from those of the bulk materials due to the quantum confinement effect. Silver [1, 2], copper [2–4], and gold [2, 5] are among the most useful metals suited for nanoparticle preparation for optoelectronics and nonlinear optics. A further search for prospective materials in clustered form, and their preparation and application are of considerable importance nowadays. In particular, the formation of cluster-containing laser plumes using relatively long (nanosecond and picosecond) laser pulses seems to be an attractive area of study due to the availability of such lasers in many laboratories. Knowledge of the conditions when clusters are present in plasma plumes opens the door for the study of various properties of these media, in particular their high-order nonlinear optical properties, which allow the creation of coherent sources of efficient HHG of laser radiation.

4.1. Overview

Nanoparticle formation during laser ablation of targets has previously been described as a process of short excitation of electronic gas and transfer of this energy to the atomic cell with further aggregation processes, which continue during evaporation of the material [6]. In the case of bulk target ablation, attention is focused on creation of the conditions when laser energy is accumulated for a short period in a small area to maintain the conditions of nonequilibrium heating. In that case, the extremely heterogeneous conditions help to create nanoparticles in the small heated areas of samples. Ablation-induced nanoparticle formation in laser plumes has been carefully documented in multiple experiments using femtosecond laser pulses [7, 8]. One can maintain the conditions such that the aggregated atoms do not disintegrate during evaporation from the surface. The analysis of the aggregation state of evaporated particles was carried out by various techniques. Among them, time-resolved emission spectroscopy, CCD camera imaging of plasma plumes, Rayleigh scattering, laser-induced fluorescence, etc., have shown ability in defining the presence of nanoparticles in the laser plumes.

Currently, nanoparticles, fullerene clusters, and carbon nanotubes (CNTs) with various sizes and shapes are commercially available from many manufacturers and they can be attached to surfaces and then evaporated using the laser ablation technique. For that use, one has to carefully define the optimum laser intensity, pulse duration, and focusing conditions for heating of the nanoparticles until they evaporate from the targets [9–13]. The history of the ablated nanoparticles can be difficult to analyze using the above techniques due to restrictions in identification of the clusters in plasma plumes at moderate heating of the targets. In that case the comparison of the size characteristics of the initial nanoparticles and deposited debris becomes a versatile approach for definition of the changes of nanoparticle morphology during laser ablation. Another indirect method is analysis of the harmonic spectra generated in plasmas containing nanoparticles and atoms/ions of the same origin.

It is well accepted that, when a solid target is ablated by laser radiation, the ablated material consists of atoms and clusters. These atoms and clusters tend to aggregate during the laser pulse or soon afterwards, leading to the formation of larger clusters. The reported results (see for example [14]) also indicate that the ablation processes on the picosecond and femtosecond timescales differ considerably compared to the nanosecond one. In addition to earlier experimental observations, several theoretical studies have suggested that rapid expansion and cooling of solid-density matter heated by short laser pulses may result in nanoparticle synthesis via different mechanisms. Heterogeneous decomposition, liquid phase ejection and fragmentation, homogeneous nucleation and decomposition, and photomechanical ejection are among the processes that can lead to nanoparticle production [15–17]. Short pulses, unlike the nanosecond pulses, do not interact with the ejected material, thus avoiding complicated secondary laser interactions. Further, short pulses heat a solid to higher temperature and pressure than do longer pulses of comparable fluence, since the energy is delivered before significant thermal conduction can take place.

Pulsed laser deposition techniques in rare gas ambients have been used for nanoparticle preparation, multicomponent thin-film deposition, and CNT synthesis. As a result of frequent collisions of ablated particles with gas atoms in a plume, the particles cool down and form nanoparticles. The development of nanoparticle synthesis in gas conditions and the characterization techniques that make possible the control of nanoparticle features within a few nanometers have attracted renewed interest in the production of metal nanoparticles, as this opens up the possibility of taking advantage of their special properties for the development of applications such as new catalysts, tunnel resonance resistors, or optical devices. The search for the optimal gas pressure for nanoparticle formation during laser ablation of solids thus remains an open issue.

It is known that metal ablation in air is less efficient than that in vacuum due to re-deposition of the ablated material. The influence of the surrounding gas on the conditions of cluster formation during laser ablation of the metals by short laser pulses has previously been analyzed for only two conditions: when the target was placed in vacuum or in ambient air. The ablation rates in vacuum can be calculated using a thermal model, as well as from basic optical and thermal properties. It is of interest to analyze the influence of the concentration of the surrounding gas on cluster formation. One can study this process using noble gases of different atomic numbers at pressures varying between 10^{-2} torr (i.e., vacuum conditions) and 760 torr (i.e., atmospheric pressure). It would be interesting to analyze whether there is a threshold on the gas pressure scale above which the conditions for nanoparticle formation are spoiled.

The use of nanoparticles for efficient conversion of the wavelength of ultrashort laser light towards the XUV range through harmonic generation seems an attractive application of cluster-containing plasmas. Note that earlier observations of HHG in nanoparticles were limited to using the exotic gas clusters formed during fast cooling of atomic flows from gas jets [18–22]. One can assume that there are difficulties in the definition of the structure of such clusters and the ratio between nanoparticles and atoms/ions in the gas flow. The characterization of gas-phase cluster production is currently being improved using sophisticated techniques (e.g., control of nanoparticle mass and spatial distribution, see the review [23]). In the meantime, plasma nanoparticle HHG has demonstrated some advantages over gas cluster HHG [24]. The application of commercially available nanopowders allows precise defininition of the size and structure of these clusters in the plume. The laser ablation technique has made possible the predictable manipulation of plasma consistency, which led to the creation of laser plumes containing mainly nanoparticles with known spatial structure. The latter allows the application of such plumes in nonlinear optics, X-ray emission from clusters, deposition of nanoparticles with fixed parameters on substrates for the semiconductor industry, production of nanostructured and nanocomposite films, etc.

Other nanostructures that have attracted attention previously are the fullerenes and CNTs. Recently, application of the laser ablation technique has allowed the creation of relatively dense C_{60}-rich plasma ($\sim 5 \times 10^{16} \, cm^{-3}$), in stark contrast with oven-based heating methods ($< 10^{14} \, cm^{-3}$). Efficient broadband HHG has been demonstrated in C_{60}-rich plasmas [25]. The changes in fundamental wave characteristics allowed dramatic manipulation of the harmonic spectrum and intensity at well-defined conditions of C_{60}-containing plasma. The application of CNT-containing plasma for efficient harmonic generation has also been proposed [26]. For these purposes, one has to create CNT-containing plasmas where the presence of nanotubes can be proven by indirect methods. More details on HHG studies in fullerenes and CNTs will be presented in the following sections.

Here we describe the studies of conditions when the plasma produced on the surface of targets contains nanoparticles, clusters, and nanotubes. These studies show that nanoparticle formation in plasma plumes can be accomplished using relatively long (subnanosecond) laser pulses under tight focusing conditions. Laser ablation of various metals in vacuum at tight and weak focusing conditions of Ti:sapphire laser radiation leads to the synthesis of 60 nm nanoparticles. Nanoparticle formation during laser ablation of metals strongly depends on the concentration of the surrounding gas. The studies show that destruction of nanoparticle formation is attributed to the negative influence of surrounding gas particles on ablated particle aggregation. We discuss the morphology of ablated nanoparticles after their laser-induced deposition on various substrates, thus confirming the presence of nanoparticles in the plasmas under appropriate ablation conditions. At moderate laser intensity of subnanosecond pulses on the surface of nanoparticle-containing materials ($<5 \times 10^9$ W cm^{-2}), the deposited material remains approximately the same as the initial nanoparticles. The cluster-containing plasma plumes prove to be effective media for HHG of femtosecond laser pulses. Finally, we describe HHG in various plasmas containing nanoparticles, clusters, and nanotubes.

4.2. Ablation of Metal Nanoparticles

Various commercially available nanoparticles were glued onto glass substrates by mixing them with a drop of superglue. Bulk materials of the same origin as the nanoparticle powders were used to compare the ablation properties of these targets. Targets containing silver, gold, platinum, ruthenium, and palladium nanoparticles (Wako Industries, Japan) were also used in these studies for creation of clustered plumes. The latter nanoparticles were purchased in the form of suspensions and were dried on the surfaces of glass substrates prior to laser ablation. More details on target preparation are presented below in the corresponding sections.

The laser ablation was carried out at different laser intensities at the surface of the nanoparticle-containing targets. The laser intensity was maintained at conditions such that the size characteristics of the deposited material remained intact with regard to the initial morphology of the nanoparticles. The intensity of the subnanosecond ($t = 300$ ps) pulse at which these conditions were satisfied was in the range between 3×10^9 and 1×10^{10} W cm^{-2}. Under these conditions, the size of the nanoparticles deposited on the substrates during laser ablation (Fig. 4.1b) was close to the size of those glued on the surface of the targets (Fig. 4.1a). The increase of laser intensity above a certain level

(a)

(b)

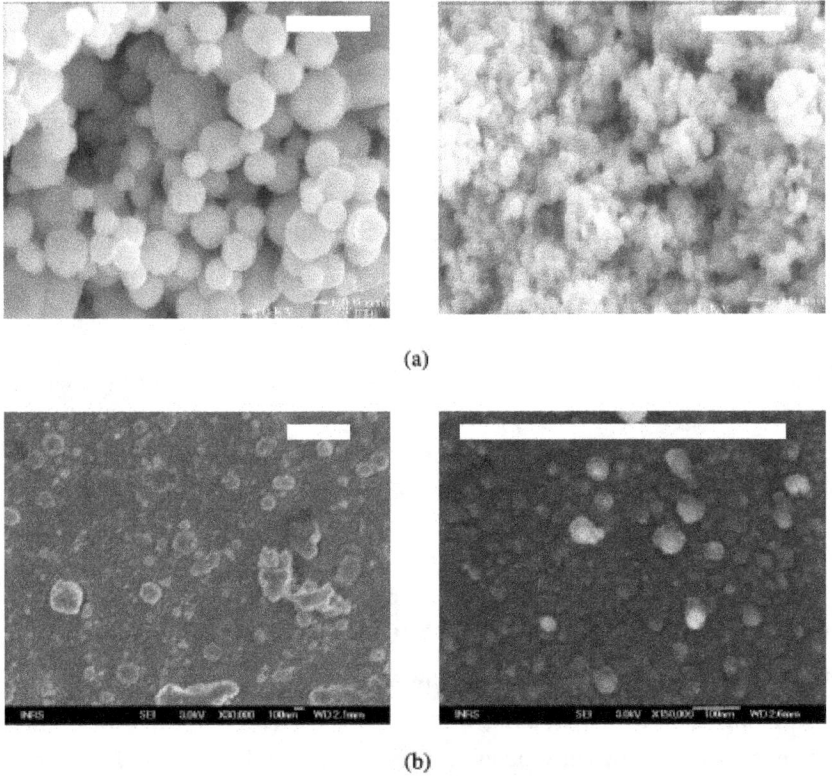

Fig. 4.1 Scanning electron microscope (SEM) images of silver (left) and copper (right) nanoparticle powders obtained (a) before ablation and (b) from the deposited substrates. The length of the scale bars is 500 nm. The powders were glued with superglue onto glass substrates. The ablation was accomplished at vacuum conditions during 50 laser shots using 300 ps heating pulses at an intensity of $I_{pp} = 7 \times 10^9$ W cm^{-2} corresponding to maximum HHG conversion efficiency from these plasma media. The ablated material was deposited onto glass substrates placed at a distance of 40 mm from the target surface.

$(I_{pp} \approx 3 \times 10^{10}$ W cm^{-2}, laser fluence 4 J cm$^{-2})$ led to considerable disintegration or aggregation of nanoparticles on the target surface. This was followed by the appearance of chaotic drops of aggregates on the substrate surface.

In most cases, the SEM images of deposited nanoparticles revealed that they remain approximately the same as the initial powders when optimal ablation of nanoparticle-containing targets was maintained. At the same time, the broader wings of size distribution observed in the histograms point to the appearance of both small and large nanoparticles due to partial aggregation and disintegration of some nanoparticles. The sizes of deposited silver clusters were in the range of 30–150 nm. The same can be said for the gold (40–180 nm), copper (30–80 nm), and other deposited clusters.

Analogous results were obtained in the case of dried nanoparticle suspensions. The initial sizes of nanoparticles (silver: 6 nm, gold: 12 nm, platinum and ruthenium: 3 nm, palladium: 5 nm) did not change considerably during drying of the suspensions, since they were protected against aggregation. Laser ablation and deposition of these nanoparticle-containing films on nearby substrates at moderate laser intensities ($I_{pp} = (3 - 8) \times 10^9\,\text{W cm}^{-2}$) led to the appearance of deposited nanoparticles possessing analogous morphology.

The presence of the original nanoparticles on the deposited substrates confirms their presence in the laser plumes. The absorption spectra of the deposited materials also confirm the presence of nanoparticles, since these spectra demonstrate SPR-induced absorption bands (Fig. 4.2a).

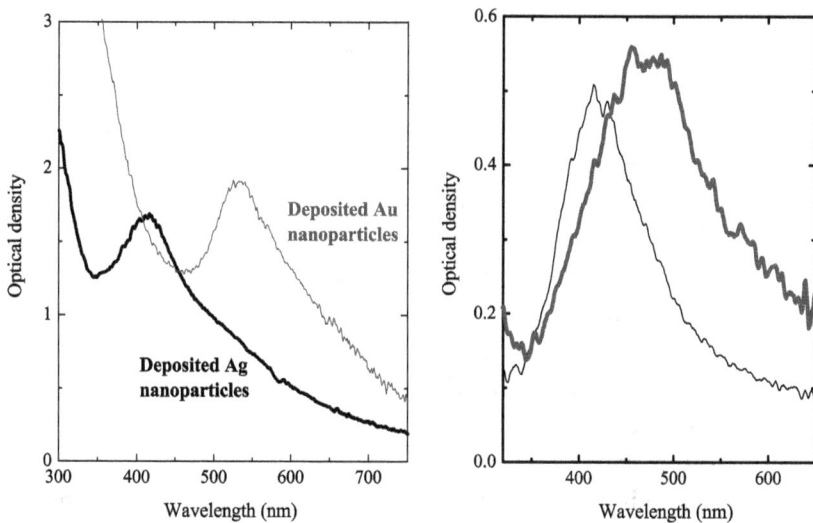

Fig. 4.2 (a) Absorption spectra of deposited silver (thick curve) and gold (thin curve) nanoparticles during ablation of nanoparticle powder-containing tagets. (b) Absorption spectra of the silver films deposited at different intensities of the heating pulse (thin curve: $I_{pp} = 1 \times 10^{12}\,\text{W cm}^{-2}$, thick curve: $I_{pp} = 6 \times 10^{12}\,\text{W cm}^{-2}$) during ablation of the bulk silver target. In these experiments, which were carried out at vacuum conditions, the 800 nm, 210 ps laser pulses irradiated the bulk target for 180 s at 10 Hz pulse repetition rate. The ablated material was deposited on glass substrates placed 50 mm away from the target.

4.3. Ablation of Bulk Metals

Here we discuss laser ablation experiments using different laser sources, when nanoparticle formation during ablation of bulk targets was confirmed by both SEM and optical absorption analysis of deposited materials. The ablation of various bulk materials was carried out in vacuum using 210 ps pulses. Uncompressed pulses from a Ti:sapphire laser were focused on a bulk target under two regimes of focusing. In the first case (referred to as tight focusing), the intensity of the laser radiation was in the range of $5 \times 10^{12} \, \mathrm{W \, cm^{-2}}$, and in the second case (referred to as weak focusing), the intensity was considerably lower ($9 \times 10^{10} \, \mathrm{W \, cm^{-2}}$). Silver and chromium were used as the targets. Float glass and GaAs wafers were used as the substrates, which were placed at a distance of 50 mm from the targets.

Figure 4.2b presents the absorption spectra of the silver films deposited on float glass substrates for the tight and weak focusing conditions. A variation of the position of the absorption peak of silver deposition was observed, which depended on the conditions of focusing of the laser pulses on the bulk target. In particular, in these cases, the peaks of SPR were centered in the range of 440 nm ($I_{pp} = 1 \times 10^{12} \, \mathrm{W \, cm^{-2}}$) and 490 nm ($I_{pp} = 6 \times 10^{12} \, \mathrm{W \, cm^{-2}}$). In the case of the deposition of silver films at lower excitation of the bulk target ($I_{pp} < 5 \times 10^{11} \, \mathrm{W \, cm^{-2}}$), no absorption peaks were observed in this region, indicating the absence of nanoparticles.

The SEM studies of the structural properties of the deposited films showed that, in the tight focusing conditions, these films contain a lot of nanoparticles of different sizes (Fig. 4.3a). In the weaker focusing conditions, the concentration of nanoparticles was considerably smaller than for the tight focusing conditions (Fig. 4.3b). The mean size of particles produced during ablation of the bulk chromium target was estimated to be 70 nm. The same behavior was observed for other targets. These studies showed that the material of the target does not play a significant role in the formation of nanoparticles in the case of laser ablation using 210 ps laser pulses at tight focusing conditions.

Further studies on the characteristics of nanosized structures of the deposited materials were carried out using atomic force microscopy (AFM). The AFM measurements were carried out in noncontact mode under an ambient environment. The mean size of silver nanoparticles deposited at tight focusing conditions was 60 nm. In contrast to this, the AFM images obtained from

(a)

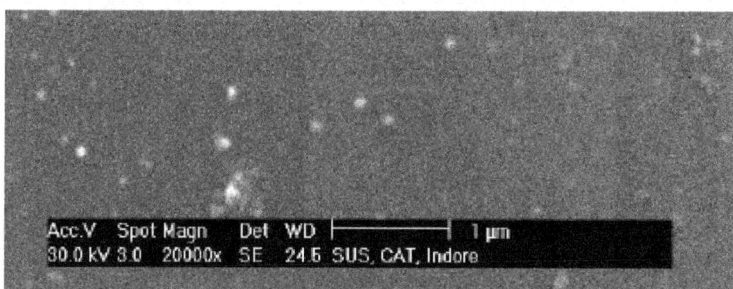

(b)

Fig. 4.3 SEM images of chromium nanoparticles deposited on a GaAs wafer placed at a distance of 50 mm from the ablation area at (a) $I_{pp} = 3 \times 10^{12} \, W \, cm^{-2}$ and (b) $I_{pp} = 7 \times 10^{11} \, W \, cm^{-2}$ during ablation of a bulk chromium target. The experiments were carried out at vacuum conditions; 210 ps laser pulses irradiated the chromium target for 200 s at 2 Hz pulse repetition rate.

the deposited films prepared at weak focusing conditions showed a considerably smaller number of nanoparticles. The same difference in AFM pictures was observed in the case of indium deposition under the two focusing conditions.

4.4. Overview of Early Studies of Harmonic Generation in Cluster-Containing Media

Various mechanisms allow us to understand the peculiarities of the interaction of strong laser pulses with an atomic/ionic medium, which in particular has led to the frequency conversion of laser radiation in the short wavelength range [27–29]. In the meantime, the enhanced nonlinear optical response of nanostructured materials induced by quantum confinement has attracted much interest, with novel applications in optoelectronics, optical switches and limiters, as well as in optical computers, optical memory, and nonlinear spectroscopy. High values of third-order nonlinear susceptibility, especially near the surface plasmon resonances of nanostructured materials, are the trademark of these media, in particular metal nanoparticles.

Clusters subject to intense laser pulses produce strong low-order nonlinear optical responses (e.g., nonlinear refraction and nonlinear absorption), as well as emitting coherent XUV radiation through harmonic generation [29, 30]. These studies have shown that one can expect an improvement of HHG efficiency by switching to cluster media. Previous studies of gas HHG from such objects were limited to exotic nanoclusters (argon, xenon), which are formed in high-pressure gas jets due to rapid cooling by adiabatic expansion.

Only a few gas HHG studies using 10^3 to 10^4 atoms/cluster media (with cluster sizes between 2 and 8 nm) have been reported [18, 20, 21], while some theoretical simulations predicted a growth of harmonic intensity with regard to monatomic media [21, 31–34]. The controlled enhancement of low-order harmonics from 6×10^5 noble-gas clusters was discussed in [35]. Clustered plasma was proposed as a nonlinear medium in which both the phase matching [31, 36] and resonantly enhanced growth of the nonlinear susceptibility [37, 38] could be achieved at selected cluster sizes and densities. However, no experimental evidence of HHG had been reported until recently, when cluster-containing laser plumes were used. One idea is to use commercially available clusters in pump-probe HHG experiments by applying laser ablation of targets containing nanoparticles.

We discuss studies of the high-order nonlinear optical properties of various nanoparticles (silver, gold, palladium, platinum, ruthenium, barium titanate ($BaTiO_3$), and strontium titanante ($SrTiO_3$)) related to HHG. Comparison of the nanoparticle nonlinearities with the high-order nonlinearities of plasma prepared on the surface of bulk targets of the same origin is also discussed.

4.5. Application of Cluster-Containing Plasma for Efficient HHG

A few examples of applications of deposition analysis techniques are presented below, which enable us to detect the presence of clusters in plumes, for studies of the high-order nonlinear optical properties of nanosized species in laser plasmas. Plumes containing various metal nanoparticles were used for HHG of the femtosecond radiation propagating through the laser plasmas. The harmonics were generated effectively at the conditions when the presence of nanoparticles in the plumes was confirmed by morphological analysis of the debris [39]. The high-order harmonics were observed in all the studied nanoparticle-containing laser plasmas. Focusing of femtosecond radiation in front of or after the laser plume optimized the harmonic yield. Focusing inside the plasma area led to a decrease of harmonic efficiency due to over-ionization of nanoparticles by femtosecond pulses and the appearance of additional free electrons. The latter led to phase mismatch of the HHG, which has previously been reported in the case of over-excitation of bulk targets [30].

The details of HHG from the ablated nanoparticles can be found elsewhere [9, 10, 40, 41]. Here we show the characteristic peculiarity of this process — the enhancement of low-order harmonic yield from 100 nm silver nanoparticles compared with atoms and ions of the same material (Fig. 4.4) [13].

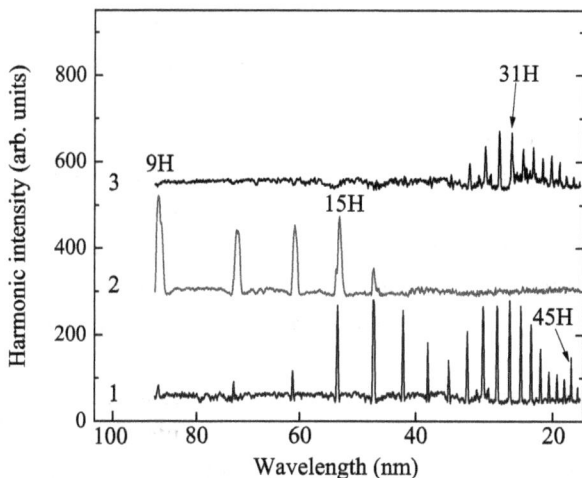

Fig. 4.4 Harmonic spectra obtained from plasmas produced on the surfaces of (1) bulk silver, and (2, 3) a silver nanoparticle-containing target. The laser intensities of the heating subnanosecond pulses on the target surfaces were (1) $3 \times 10^{10}\,\mathrm{W\,cm^{-2}}$, (2) $7 \times 10^9\,\mathrm{W\,cm^{-2}}$, and (3) $2 \times 10^{10}\,\mathrm{W\,cm^{-2}}$. The intensity of the femtosecond pulse in the plasma plume was $4 \times 10^{14}\,\mathrm{W\,cm^{-2}}$. The delay between the subnanosecond and femtosecond pulses was maintained at 20 ns. The femtosecond pulse propagated through the plasma 100 μm above the target surface. Adapted from [13] with permission from American Institute of Physics.

One can see the six- to twenty-fold enhancement of low-order (9th, 11th, and 13th) harmonic generation efficiency in the silver nanoparticle-containing plasma plumes (Fig. 4.4, curve 1) compared with the same process in the plasmas produced on the surface of bulk silver (Fig. 4.4, curve 2). The increase of heating pulse intensity from $7 \times 10^9 \, \mathrm{W \, cm^{-2}}$ (Fig. 4.4, curve 2) to $2 \times 10^{10} \, \mathrm{W \, cm^{-2}}$ (Fig. 4.4, curve 3) led to disintegration of nanoparticles and the corresponding disappearance of strong low-order harmonics, while the appearance of higher-order harmonics confirmed the involvement of atomic/ionic particles in the laser frequency conversion.

One can assume that, in the case of small nanoparticles, the ejected electron, after returning back to the parent particle, can recombine with any of the atoms in the nanoparticle due to the enhanced cross section of recombination with the parent particle compared with a single atom, which considerably increases the probability of harmonic emission in the former case. Thus, the enhanced cross section of recombination of an accelerated electron with the parent particle compared to a single atom can be a reason for the observed enhancement of HHG yield from the nanoparticle-containing plume compared with the atom/ion-containing plasma.

The HHG from the interaction of *in-situ* produced nanoparticles with intense ultrashort laser pulses has also been studied [42]. The interaction of a subnanosecond pulse with a silver target at intensity of $\sim 1 \times 10^{13} \, \mathrm{W \, cm^{-2}}$ generates the nanoparticles. The mean size of silver nanoparticles was 30 nm. High-order harmonics were generated through the interaction of *in-situ* produced nanoparticles with ultrashort laser pulses. The spectrum of HHG from *in-situ* produced nanoparticles was compared with the HHG spectrum from bulk silver plumes and 9 nm silver nanoparticles glued onto the target (Fig. 4.5).

The intensities of the 9th to 15th harmonics were lower than the 17th harmonic in the case of HHG from atom/ion-containing silver plumes. Also, the intensity of harmonics from the plumes created on the target coated by silver nanoparticles decreases slowly from the 9th harmonic to the 17th harmonic. Comparison of the HHG spectral characteristics from *in-situ* produced nanoparticles with those from bulk silver and glued silver nanoparticles indicates that HHG in that case is from silver nanoparticles rather than from silver atoms and ions. As the intensity of the heating pulse on the bulk silver surface was increased from $\sim 10^{10} \, \mathrm{W \, cm^{-2}}$ to $\sim 8 \times 10^{11} \, \mathrm{W \, cm^{-2}}$,

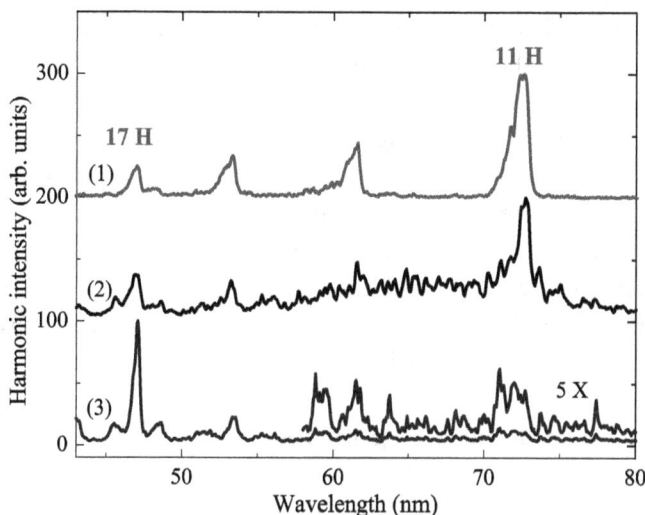

Fig. 4.5 Comparison of the HHG spectra from (1) silver nanoparticles glued onto the target, (2) *in-situ* produced silver nanoparticles, and (3) silver plasma without the nanoparticles. Reprinted from [42] with permission from American Physical Society.

the HHG spectra gradually reduced and completely vanished. Then, at $8 \times 10^{12} - 1 \times 10^{13}\,\mathrm{W\,cm^{-2}}$, HHG suddenly reappeared, as the radiation was focused at tight focusing conditions. The intensity of HHG radiation from *in-situ* produced nanoparticles is lower than that from plasma produced on the silver nanoparticle-coated surface. However, the intensity of HHG could be further enhanced by improving the methods for nanoparticle formation.

4.6. Peculiarities of HHG in Nanoparticle-Containing Plasmas

The first HHG experiments using 100 nm $BaTiO_3$ and 40 nm $SrTiO_3$ nanoparticles glued onto different substrates were reported in [9]. The ablation of these media was performed using a low intensity, 210 ps pulse, which was followed by propagation of a femtosecond pulse through the plasma to generate the harmonics. As was normally done in previous harmonic experiments using bulk targets, a single position on the surface of the nanoparticle-containing substrate was irradiated without moving the target. In experiments with bulk targets, one can generate harmonics with good shot-to-shot reproducibility with this method. However, the studies using silver nanoparticle powder-containing targets showed the opposite tendency [43]. The first shot on a fresh target surface produced strong harmonic generation, which was much stronger than that observed using bulk silver targets for the same conditions. This was followed by a steep decrease in the harmonic yield after a few shots at the same target position. A tendency for harmonic intensity to decrease after several consecutive irradiations was attributed to the fast evaporation of a thin layer of nanoparticles from the surface after a few shots. Because of this, thick layers of $BaTiO_3$ and $SrTiO_3$ nanoparticle powders were prepared to carry out the experiments for a relatively long period without changing the conditions of particle evaporation.

The presence of rather large nanoparticles diminishes the role of the size-related effect on the nonlinear optical response of the plume containing $BaTiO_3$ and $SrTiO_3$ multi-atomic particles. In the case of $BaTiO_3$ nanoparticle-containing laser plumes, harmonics up to the 39th order were generated (Fig. 4.6a). The harmonic cutoff for the plume created on the surface of the $BaTiO_3$ bulk crystal (41st harmonic, Fig. 4.6b, curve 1) was approximately the same as that obtained from the plasma created on the $BaTiO_3$ nanoparticle-containing substrates. An increase of heating pulse intensity on the crystal surface above some limit led to considerable growth of the plasma emission (Fig. 4.6b, curve 2). The harmonic conversion efficiencies at the plateau range were estimated to be in the range of 4×10^{-6} for both plasmas, assuming the same experimental conditions for the present studies and the investigations of HHG from a bulk silver plume, when the absolute calibration of harmonic yield was established [44]. The equality of harmonic cutoffs and efficiencies pointed to the absence of the influence of quantum confinement-induced processes during HHG from the 100 nm nanoparticle-containing plumes compared with the monoparticle-containing medium of the same origin. Furthermore, one has to take into account that, in the case of $BaTiO_3$, the influence of quantum confinement can be expected for particles with sizes smaller than 16 nm [36].

The same tendency was observed in the case of HHG from the 40 nm $SrTiO_3$ nanoparticle-containing laser plumes. While the harmonic cutoff (39th harmonic) was not compared with that from the bulk $SrTiO_3$, higher

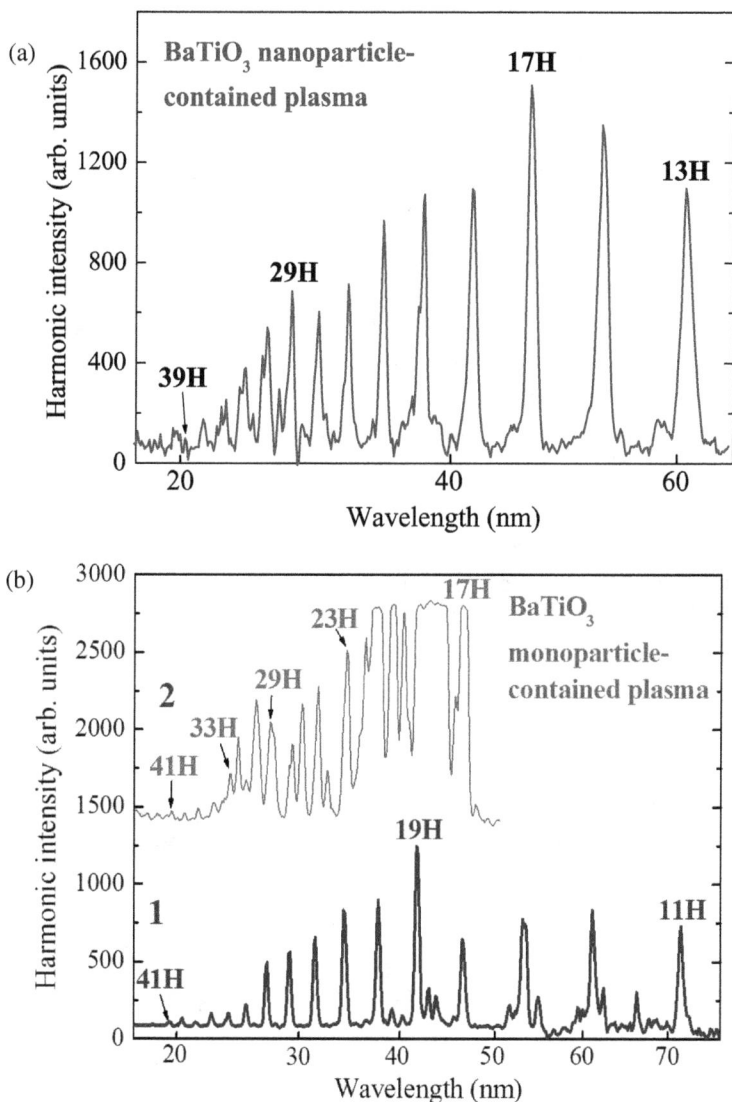

Fig. 4.6 High-order harmonic spectra obtained from plasmas produced on the surfaces of (a) $BaTiO_3$ nanoparticle-containing substrates (heating pulse intensity $8 \times 10^9 \, W \, cm^{-2}$) and (b) $BaTiO_3$ bulk crystal. Curves 1 and 2 on (b) correspond to heating pulse intensities of 3×10^{10} and $5 \times 10^{10} \, W \, cm^{-2}$, respectively. Reprinted from [9] with permission from Optical Society of America.

conversion efficiency was observed in the case of $SrTiO_3$ nanoparticle-containing laser plumes compared with the case of $BaTiO_3$ nanoparticle-containing plasma. The ratio of the harmonic yields from these two cluster-containing plumes was in the range of four, and depended on the heating pulse intensity at the nanoparticle-containing surface. Note the equality of nanoparticle concentration for the $SrTiO_3$ and $BaTiO_3$ compounds glued onto the substrates. It was initially verified that the harmonics generated from these substrates and the glue itself (drop of glue, tape, and glass), without nanoparticles, were negligible compared with those from the $BaTiO_3$ and $SrTiO_3$ nanoparticle-containing plasmas.

Comparison of HHG from the plasma produced on the surface of bulk silver and the target containing silver nanoparticles of different sizes was analyzed in [10]. The HHG from the plasma created on the surface of bulk silver targets has been optimized using Ti:sapphire lasers generating pulses of different duration (150 fs [44], 48 fs [45], and 35 fs [46]). Silver plasma still remains the best from the point of view of the highest conversion efficiency at the plateau region (8×10^{-6} [47]). Previous work revealed that, at the "optimal plasma" conditions, which means the best output characteristics of generated harmonics, no nanoparticles existed in the laser plume produced on the surface of the bulk target [48].

The initial experiments were carried out using 110 nm silver powder nanoparticles. Figure 4.7 shows three sets of comparative measurements of the harmonic yield from silver plasmas containing monoatoms and monoions (thin lines) and nanoparticles (thick lines). These measurements were performed at three different delays between the heating picosecond pulse and the probe femtosecond pulse. At small delay (5 ns), the harmonic output from the nanoparticle-containing plume exceeded that from the monoparticle plasma (Fig. 4.7a). Another pattern was observed for longer delay (17 ns). The harmonics from the plasma produced on the surface of bulk silver prevailed over HHG from the nanoparticle-containing medium (Fig. 4.7b). This feature remained unchanged up to the maximum delays (150 ns) used in these studies (see also Fig. 4.7c showing the harmonic spectrum at 88 ns delay).

It was found that HHG from the plume containing silver multi-atomic particles became more efficient at considerably smaller pulse intensities ($4 \times 10^9 \, W \, cm^{-2}$) compared to the case of the bulk silver target ($2 \times 10^{10} \, W \, cm^{-2}$). This tendency, which was also observed in the cases of $BaTiO_3$ and $SrTiO_3$

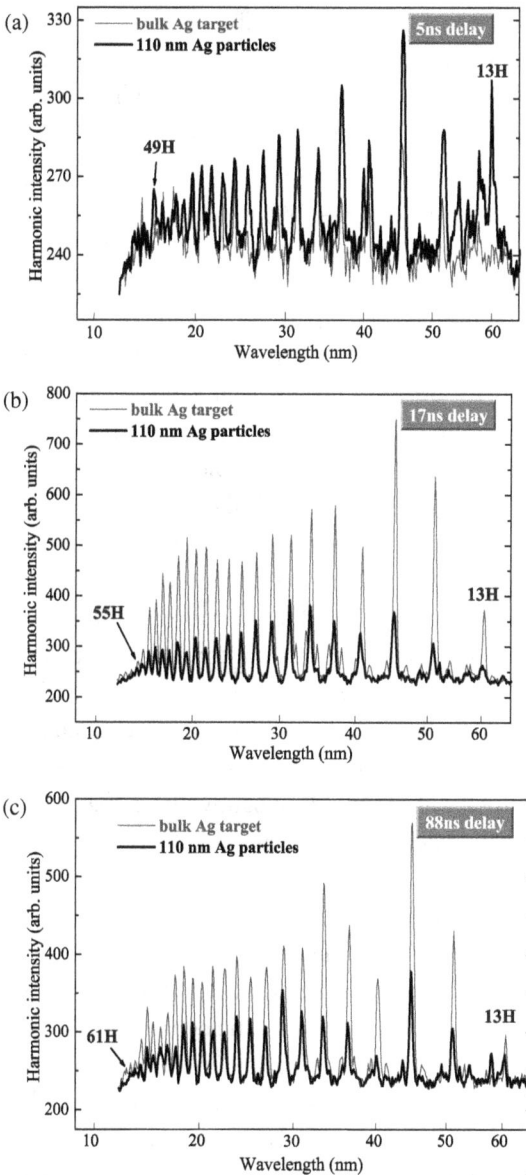

Fig. 4.7 Harmonic spectra from plasma containing 110 nm silver particles (thick lines) and plasma created on a bulk silver surface (thin lines) at (a) 5 ns, (b) 17 ns, and (c) 88 ns delay between the subnanosecond heating pulse and the femtosecond probe pulse. Heating pulse intensity in these experiments was maintained at the level of 1×10^{10} W cm^{-2}. Reprinted from [10] with permission from Institute of Physics.

nanoparticle-containing targets, was attributed to the lower ablation threshold and lower cohesive energy of the host substrate that contains nanoparticles compared to those of bulk material. The presence of rather large nanoparticles caused by a very broad size distribution makes it difficult to predict the influence of quantum confinement on the nonlinear optical response of the plume containing silver multi-atomic particle powder.

For silver nanoparticle powder-containing laser plumes, harmonics as high as 55th order were achieved (Figs. 4.7b and 4.7c). The harmonic cutoff for the plume created on the surface of the bulk silver target was approximately the same as that obtained from the plasma created on the silver powder-containing substrates. The prevailing of the harmonic yield from the nanoparticle-containing medium over the monatomic one in the case of short delays can be attributed to the more favorable conditions of evaporation of the nanoparticle-containing medium, which allowed higher particle concentration to be achieved at the area of femtosecond beam propagation compared to the laser ablation of the bulk target at the initial stages of ablation.

The absence of enhancement of the harmonic yield from the nanoparticle-containing plasma (at equal experimental conditions, when the optimal silver plasma was generated) was probably caused by relatively high sizes of the silver nanoparticles. The application of smaller nanoparticles can reveal new features during HHG from such structures. In particular, tuning to the resonances by applying chirped pulses of narrow-bandwidth radiation has recently been demonstrated in the case of plasma containing GaN nanoparticle powder [47] (see also Section 3.3).

Further studies have revealed that the application of smaller-sized nanoparticles prepared as a thick layer allowed the enhancement of harmonic yield from the nanoparticle-containing plume. For this purpose, small palladium, gold, platinum, and ruthenium nanoparticles dried on the surfaces of various substrates were used [11]. For 9 nm gold nanoparticle-containing laser plumes, harmonics up to 33rd order were observed (Fig. 4.8a, thick curve). The harmonic cutoff for the plume created on the surface of the gold bulk target was higher than that obtained from the plasma created on the gold nanoparticle-containing substrates (45th harmonic, Fig. 4.8a, thin curve). In the case of the gold bulk target, the harmonic yield at the longer-wavelength range of the plateau was a few times weaker than that obtained from the gold nanoparticle-containing plume (with an enhancement factor up to six, depending on the harmonic order). Figure 4.8b presents the results of measurements in the case

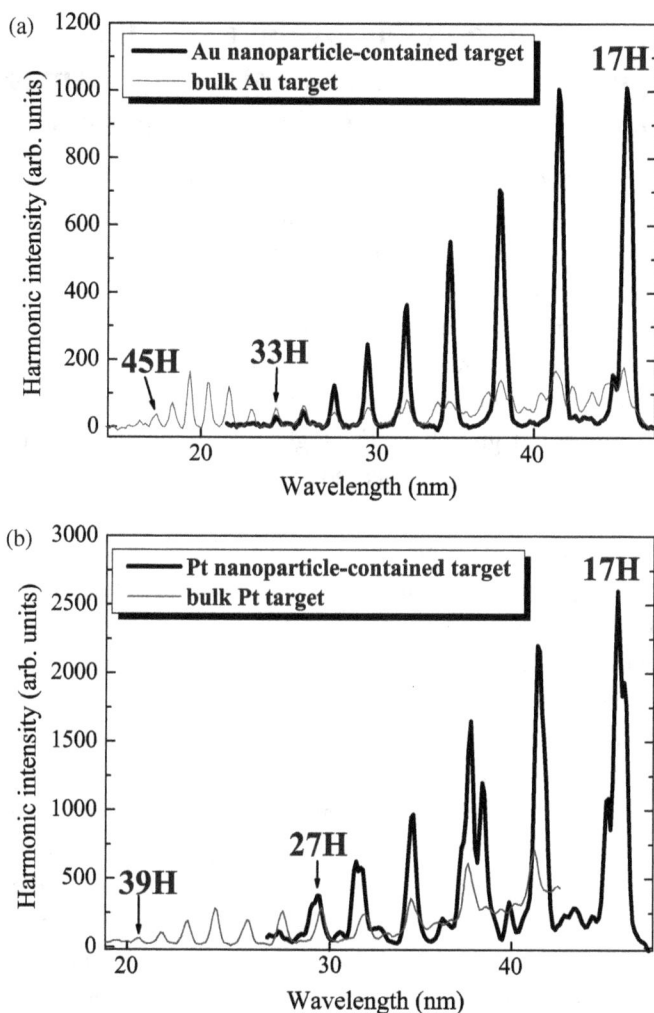

Fig. 4.8 (a) High-order harmonic spectra obtained from plasmas produced on the surfaces of gold nanoparticle-containing substrates (thick curve) and a bulk gold target (thin curve). $I_{pp} = 8 \times 10^9 \, \text{W cm}^{-2}$. (b) High-order harmonic spectra obtained from plasmas produced on the surface of platinum nanoparticle-containing substrates (thick curve) and a bulk platinum target (thin curve). Reprinted from [11] with permission from American Institute of Physics.

of 3 nm platinum multiparticle- and monoparticle-containing plasmas, which demonstrated analogous enhancement of HHG yield for the former case. The 3 nm palladium and 3 nm ruthenium nanoparticle-containing plasmas demonstrated approximately the same high-order nonlinear optical properties.

4.7. Advantages and Disadvantages of the Application of Cluster-Containing Plasmas for the Enhancement of HHG Efficiency

The studies described above have shown that no improvements in the extension of the harmonic cutoff were observed for any nanoparticle-containing plumes. At the same time, an enhancement of the harmonic yield in the low-energy plateau range in the case of small-sized nanoparticle-containing plumes was achieved. The value of the enhancement factor could be attributed to different concentrations of clusters and monoparticles in the plume at different excitation conditions, as well as to the processes related to quantum confinement-induced growth of the high-order nonlinear susceptibility of such media. Further studies are needed for a definition of the nature of these peculiarities.

The above observations give the rough pattern of the ablation and HHG from nanoparticle targets. The material directly surrounding the nanoparticles is a glue or dried polymer (polyvinylpyrrolidone (PVP) or polyethyleneimine (PEI)), which has a lower ablation threshold than the metallic materials. Therefore, the polymer starts to ablate at relatively low intensities of the heating pulse, carrying the nanoparticles with it, resulting in the lower pulse intensity required for evaporation of the target. In particular, HHG from metal multi-atomic particles started to be efficient at considerably smaller pulse intensities ($3\text{--}6 \times 10^9\,\mathrm{W\,cm^{-2}}$) compared with the case of bulk metals and crystals ($1\text{--}3 \times 10^{10}\,\mathrm{W\,cm^{-2}}$). No harmonics were generated in the case of ablation of the glue, PVP, and PEI. Repeated irradiation of the target led to melting and a change in the properties of the target containing the thin layer of nanoclusters. This was followed by a change in the conditions of the plasma plume, resulting in a reduction of the harmonic intensity after a few tens of shots.

A comparison of low-order harmonic generation using single atoms and multi-particle aggregates has previously been reported for argon atoms and clusters [18]. It was demonstrated that a medium of intermediate-sized clusters with a few thousand atoms of inert gas has a higher efficiency for generating harmonics than a medium of isolated gas atoms of the same density. The reported enhancement factor for the 3rd to 9th harmonics from gas jets was about five. In the HHG experiments with laser-ablated nanoparticles, these observations were extended towards the higher orders of harmonics when approximately the same enhancement for harmonics between the 17th and 23rd orders was achieved. These results have also shown that the dependence of HHG efficiency on the heating pulse and probe pulse intensity is much more prominent for nanoparticles than for monoparticles. Experiments on HHG using *in-situ* produced harmonics are described in [42], which showed another application of the attractive properties of clusters.

One possible reason for relatively small harmonic enhancement in the case of nanoparticles compared with the bulk material could be a change of nanoparticle structure during laser ablation. The laser intensity used to create the plume is a very sensitive factor that has to be carefully investigated before one can make statements about the integrity of nanoparticles in the plasma area after ablation. One could expect melting and aggregation of nanoparticles during the interaction of the heating pulse with the target surface. The presence of nanoparticles in the plumes in the above studies was confirmed by analyzing the spatial characteristics of the ablated material deposited on the nearby glass substrate. However, further studies of plasma components during the ablation of nanoparticle-containing targets will confirm the statement about the presence and integrity of nanoclusters in the plumes at the moment when the femtosecond pulse arrives in the area of interaction.

The above-discussed studies have shown that for larger-sized particles no improvements in either conversion efficiency or extension of the harmonic cutoff occur. To obtain maximum HHG conversion efficiency, it is essential to define the maximum tolerable particle size at which enhancement can be achieved, since an increase of the particle sizes leads to an increase of their polarizability. A large polarizability of the medium is critical for efficient harmonic generation. At the same time, an increase in cluster size leads to the appearance of impeding processes that can restrict the harmonic generation efficiency (such as the use of only surface atoms for HHG, reabsorption of harmonics, recombination of electrons from the inner parts of clusters, etc.). The problem of choosing the optimal sizes of nanoparticles for HHG experiments from plasma plumes is related to competing relevant processes, which can both decrease and increase the nonlinear optical response of a medium. An increase of nanoparticle sizes from a few nanometers to 20 nm can enhance the ionization potential of such nanostructures [33]. This will lead to an enhancement of cutoff energy. At the same time, an increase in nanoparticle size over some limit is undesirable, for the above-mentioned reasons.

The physical origin of HHG from cluster-containing laser-produced plasmas [10, 11, 47, 48] and gas clusters [18–22, 31–38] is mostly attributed to standard atomic harmonic generation, modified by the fact that in clusters the atoms are connected with each other. Another mechanism that can influence HHG efficiency from the clustered medium is the growth of the nonlinear susceptibility close to the Mie frequencies of the clusters. An example of such influence has been demonstrated through third-harmonic generation from argon clusters [35]. At the same time, attempts to achieve higher-order harmonic enhancement using the quasi-resonance conditions of the laser frequency (or its second harmonic) and Mie frequency (specifically, in the case of silver nanoparticle-containing plasma, $\lambda \approx 420$ nm) failed, probably due to insignificant growth of the nonlinear susceptibility of the medium at the wavelengths of high-order harmonics [43].

The debate is still open as to whether the generation efficiency can be further enhanced by the use of a nanoparticle-containing target. This review was aimed at providing the reader with recently reported experimental evidence on the issue. Even though many factors might be poorly controlled in the reported studies (for instance, fragmentation/condensation phenomena occurring during laser ablation of the nanoparticle-containing targets), the results presented serve as an experimental test of the role ultimately played by the state of aggregation of the matter involved in harmonic generation.

References

1. Ganeev, R.A., Ryasnyanskiy, A.I., Stepanov, A.L. et al. (2004). Saturated absorption and nonlinear refraction of silicate glasses doped with silver nanoparticles at 532 nm, Opt. Quantum Electron., 36, 949–960.
2. Ganeev, R.A., Ryasnyansky, A.I., Stepanov, A.L. et al. (2005). Application of RZ-scan technique for investigation of nonlinear optical characteristics of sapphire doped with Ag, Cu, and Au nanoparticles, Opt. Commun., 253, 205–213.
3. Ryasnyansky, A.I., Palpant, B., Debrus, S. et al. (2005). Nonlinear optical absorption of ZnO doped with copper nanoparticles in the pico- and nanosecond pulse laser field, Appl. Opt., 44, 2839–2845.
4. Falconieri, M., Salvetti, G., Cattaruza, E. et al. (1998). Large third-order optical nonlinearity of nanocluster-doped glass formed by ion implantation of copper and nickel in silica, Appl. Phys. Lett., 73, 288–290.
5. Debrus, S., Lafait, J., May, M. et al. (2000). Z-scan determination of the third-order optical nonlinearity of gold:silica nanocomposites, J. Appl. Phys., 88, 4469–4475.

6. Amoruso, S., Ausanio, G., Bruzzese, R. *et al.* (2005). Femtosecond laser pulse irradiation of solid targets as a general route to nanoparticle formation in a vacuum, *Phys. Rev. B*, 71, 033406.

7. Scuderi, D., Albert, O., Moreau, D. *et al.* (2005). Interaction of a laser-produced plume with a second time delayed femtosecond pulse, *Appl. Phys. Lett.*, 86, 071502.

8. Perrière, J., Boulmer-Leborgne, C., Benzerga, R. *et al.* (2007). Nanoparticle formation by femtosecond laser ablation, *J. Phys. D: Appl. Phys.*, 40, 7069–7076.

9. Ganeev, R.A., Suzuki, M., Baba, M. *et al.* (2008). Low- and high-order nonlinear optical properties of $BaTiO_3$ and $SrTiO_3$ nanoparticles, *J. Opt. Soc. Am. B*, 25, 325–333.

10. Ganeev, R.A., Suzuki, M., Baba, M. *et al.* (2008). High-order harmonic generation in Ag nanoparticle-containing plasma, *J. Phys B: At. Mol. Opt. Phys.*, 41, 045603.

11. Ganeev, R.A., Suzuki, M., Baba, M. *et al.* (2008). Low- and high-order nonlinear optical properties of Au, Pt, Pd, and Ru nanoparticles, *J. Appl. Phys.*, 103, 063102.

12. Elouga Bom, L.B., Ganeev, R.A., Abdul-Hadi, J. *et al.* (2009). Intense multi-microjoule high-order harmonics generated from neutral atoms of In_2O_3 nanoparticles, *Appl. Phys. Lett.*, 94, 111108 (2009).

13. Ganeev, R.A., Elouga Bom, L.B. and Ozaki, T. (2009). Application of nanoparticle-containing laser plasmas for harmonic generation, *J. Appl. Phys.*, 106, 023104.

14. Teghil, R., D'Alessio, L., Santagata, A. *et al.* (2003). Picosecond and femtosecond pulsed laser ablation and deposition of quasicrystals, *Appl. Surf. Sci.*, 210, 307–317.

15. Glover, T.E. (2003). Hydrodynamics of particle formation following femtosecond laser ablation, *J. Opt. Soc. Am. B*, 20, 125–131.

16. Jeschke, H.O., Garsia, M.E. and Bennemann, K.H. (2001). Theory for the ultrafast ablation of graphite films, *Phys. Rev. Lett.*, 87, 015003.

17. Kabashin, V. and Meunier, M. (2003). Synthesis of colloidal nanoparticles during femtosecond laser ablation of gold in water, *J. Appl. Phys.*, 94, 7941–7944.

18. Donnelly, T.D., Ditmire, T., Neuman, T. *et al.* (1996). High-order harmonic generation in atom clusters, *Phys. Rev. Lett.*, 76, 2472–2475.

19. Hu, S.X. and Xu, Z.Z. (1997). Enhanced harmonic emission from ionized clusters in intense laser pulses, *Appl. Phys. Lett.*, 71, 2605–2607.

20. Tisch, J.W.G., Ditmire, T., Frasery, D.J. *et al.* (1997). Investigation of high-harmonic generation from xenon atom clusters, *J. Phys. B: At. Mol. Opt. Phys.*, 30, L709–L713.

21. Vozzi, C., Nisoli, M., Caumes, J.-P. *et al.* (2005). Cluster effects in high-order harmonics generated by ultrashort light pulses, *Appl. Phys. Lett.*, 86, 111121.

22. Pai, C.-H., Kuo, C.C, Lin, M.-W. *et al.* (2006). Tomography of high harmonic generation in a cluster jet, *Opt. Lett.*, 31, 984–986.

23. Wegner, K., Piseri, P., Tafreshi, H.V. *et al.* (2006). Cluster beam deposition: A tool for nanoscale science and technology, *J. Phys. D: Appl. Phys.*, 39, R439–R470.

24. Ganeev, R.A. (2008). High-order harmonic generation in nanoparticle-containing laser-produced plasmas, *Laser Phys.*, 18, 1009–1015.

25. Ganeev, R.A., Elouga Bom, L.B., Abdul-Hadi, J. *et al.* (2009). High-order harmonic generation from fullerene using the plasma harmonic method, *Phys. Rev. Lett.*, 102, 013903.

26. Ganeev, R.A., Naik, P.A., Singhal, H. *et al.* (2011). High order harmonic generation in carbon nanotube-containing plasma plumes, *Phys. Rev. A*, 83, 013820.

27. Hasović, E., Milošević, D.B., Becker, W. *et al.* (2006). A method of carrier-envelope phase control for few-cycle laser pulses, *Laser Phys. Lett.*, **3**, 200–204.

28. Kramo, A., Hasović, E., Milošević, D.B. *et al.* (2007). Above-threshold detachment by a two-colour bicircular laser field, *Laser Phys. Lett.*, **4**, 279–286.

29. Shvetsov-Shilovski, N.I., Goreslavski, S.P., Popruzhenko, S.V. *et al.* (2007). Reconstruction of an arbitrarily polarized few-cycle laser pulse by two-dimensional streaking, *Laser Phys. Lett.*, **4**, 726–733.

30. Liao, H.B., Xiao, R.F., Fu, J.S. *et al.* (1997). Large third-order nonlinear optical susceptibility of Au-Al$_2$O$_3$ composite films near the resonant frequency, *Appl. Phys. B*, **65**, 673–676.

31. Tisch, J.W.G. (2000). Phase-matched high-order harmonic generation in an ionized medium using a buffer gas of exploding atomic clusters, *Phys. Rev. A*, **62**, 041802.

32. Veniard, V., Taieb, R. and Maquet, A. (2001). Atomic clusters submitted to an intense short laser pulse: A density-functional approach, *Phys. Rev. A*, **65**, 013202.

33. Vazquez de Aldana, J.R. and Roso, L. (2001). High-order harmonic generation in atomic clusters with a two-dimensional model, *J. Opt. Soc. Am. B*, **18**, 325–330.

34. Kundu, M., Popruzhenko, S.V. and Bauer, D. (2007). Harmonic generation from laser-irradiated clusters, *Phys. Rev. A*, **76**, 033201.

35. Shim, B., Hays, G., Zgadzaj, R. *et al.* (2007). Enhanced harmonic generation from expanding clusters, *Phys. Rev. Lett.*, **98**, 123902.

36. Tajima, T., Kishimoto, Y. and Downer, M.C. (1999). Optical properties of cluster plasma, *Phys. Plasmas*, **6**, 3759–3764.

37. Fomits'kyi, M.V., Breizman, B.N., Arefiev, A.V. *et al.* (2004). Harmonic generation in clusters, *Phys. Plasmas*, **11**, 3349–3359.

38. Fomichev, S.V., Zaretsky, D.F., Bauer, D. *et al.* (2005). Classical molecular-dynamics simulations of laser-irradiated clusters: Nonlinear electron dynamics and resonance-enhanced low-order harmonic generation, *Phys. Rev A*, **71**, 013201.

39. Ganeev, R.A., Elouga Bom, L.B. and Ozaki, T. (2009). Comparison of high-order harmonic generation from various cluster- and ion-containing laser plasmas, *J. Phys. B: At. Mol. Opt. Phys.*, **42**, 055402.

40. Singhal, H., Ganeev, R.A., Naik, P.A. *et al.* (2010). Study of high-order harmonic generation from nanoparticles, *J. Phys. B: At. Mol. Opt. Phys.*, **43**, 025603.

41. Ozaki, T., Elouga Bom, L.B., Abdul-Haji, J. *et al.* (2010). Evidence of strong contribution from neutral atoms in intense harmonic generation from nanoparticles, *Laser Part. Beams*, **28**, 69–74.

42. Singhal, H., Ganeev, R.A., Naik, P.A. *et al.* (2010). In-situ laser induced silver nanoparticle formation and high order harmonic generation, *Phys. Rev. A*, **82**, 043821.

43. Ganeev, R.A. (2007). High-order harmonic generation in laser plasma: A review of recent achievements. *J. Phys. B: At. Mol. Opt. Phys.*, **40**, R213–R253.

44. Ganeev, R.A., Baba, M., Suzuki, M. *et al.* (2005). High-order harmonic generation from silver plasma, *Phys. Lett. A*, **339**, 103–109.

45. Ganeev, R.A., Singhal, H., Naik, P.A. *et al.* (2007). Optimization of the high-order harmonics generated from silver plasma, *Appl. Phys. B*, **87**, 243–247.

46. Elouga Bom, L.B., Kieffer, J.-C., Ganeev, R.A. *et al.* (2007). Influence of the main pulse and prepulse intensity on high-order harmonic generation in silver plasma ablation, *Phys. Rev. A*, **75**, 033804.

47. Ganeev, R.A., Suzuki, M., Redkin, P.V. *et al.* (2007). Variable pattern of high harmonic spectra from a laser-produced plasma by using the chirped pulses of narrow-bandwidth radiation, *Phys. Rev. A*, 76, 023832.

48. Ganeev, R.A., Chakravarty, U., Naik, P.A. *et al.* (2007). Pulsed laser deposition of metal films and nanoparticles in vacuum using subnanosecond laser pulses, *Appl. Opt.*, 46, 1205–1210.

5

Application of Fullerenes for Harmonic Generation

The relatively low HHG efficiency is still an obstacle for practical application of harmonic radiation. The maximum available efficiency for harmonics in the plateau range commonly remains at a level of below 10^{-5} (except for the extremely strong 13th harmonic produced in an indium plasma plume, with an efficiency about 10^{-4} [1]). Thus, the search for new methods for improvement of HHG efficiency is an important goal of nonlinear optics. The alternative to previous approaches is to search for media possessing those properties that would allow an increase of HHG efficiency. In this connection, small-sized nanostructures are an attractive alternative, since they demonstrate local-field-induced enhancement of the nonlinear optical response of a medium. This peculiarity has been used for enhancement of low-order harmonics in the vicinity of the SPR of nanoparticles [2]. Another possible mechanism that can enhance the harmonic efficiency is an increase of the recombination cross section of accelerated electron and parent particle in the last stage of the three-step mechanism of HHG. More details were discussed in previous chapters.

In this connection, fullerenes can be considered as an attractive nonlinear medium for HHG. Their relatively large sizes and broadband SPR in the XUV ($\lambda_{SPR} \cong 60$ nm with 10 nm full width at half maximum) allowed the first demonstration of efficient HHG from fullerenes near their SPR [3]. The application of a laser ablation technique allowed the creation of relatively

113

dense C_{60}-rich plasma ($\sim 5 \times 10^{16}$ cm^{-3}), in stark contrast with the density $\leq 10^{15}$ cm^{-3} obtained in oven-based heating methods for production of fullerene beams.

C_{60} can be considered as a prospective nonlinear medium because (i) it is highly polarizable, ~ 80 Å^3 [3]; (ii) it is stable against fragmentation in intense laser fields due to its very large number of internal degrees of freedom leading to the fast diffusion of excitation energy; (iii) it exhibits giant plasmon resonance at the photon energy of ~ 20 eV [4]; (iv) it has large photoionization cross sections [4, 5]; and (v) multi-electron dynamics is known to influence the ionization and recollision [6, 7] that are central to the HHG process. The saturation intensities of different charge states of C_{60} are higher than those of isolated atoms of similar ionization potential [6, 8].

The nonlinear optical parameters responsible for low-order (i.e., second and third) harmonic generation in fullerenes were analyzed in [9] and [10] respectively. High values of nonlinear optical susceptibilities were obtained in those studies ($\chi^{(3)}$ ($-3\ \omega$; ω, ω, ω) $= 2 \times 10^{-10}$ esu and $\chi^{(3)}$ ($-2\omega; \omega, \omega, 0$) $= 2.1 \times 10^{-9}$ esu for the C_{60} films at the wavelength of 1064 nm). Previous studies of fullerenes have also demonstrated generation of the fifth harmonic [11]. Note the absence of reports on higher-order harmonics in fullerenes until the studies where the application of laser ablation allowed the production of plasma plumes containing considerable amounts of fullerene particles for efficient conversion of short laser pulses (i.e., of a few tens of femtoseconds) in the XUV range [3, 12–14].

Below, we present an analysis of experimental and theoretical results related to studies of the high-order nonlinear optical properties of C_{60}. We discuss the results of systematic studies of HHG in C_{60}-rich laser-produced plasma under various plasma conditions and laser parameters. Specifically, we describe (i) the enhancement of harmonic yield near SPR, and (ii) the extension of the harmonic cutoff, using different driving wavelengths and polarization and by optimizing the delay between the picosecond heating pulse and the femtosecond probe pulse. We also analyze the morphology of clusters before and after ablation to define the optimal conditions of excitation of the C_{60}-containing targets. We show that the enhancement of HHG near SPR of C_{60} is independent of the driving laser wavelength. We demonstrate further enhancement of HHG efficiency in C_{60}-rich plasmas by using the two-color driving pulse technique previously used for gas HHG. A two-color pump using fundamental and second-harmonic radiation at different experimental conditions of both the fullerene-containing laser plasma plume and overlapping probe pulses allowed the manipulation of the harmonic spectrum and intensity at well-defined conditions of C_{60}-containing plasma. The conversion efficiency for the odd and even harmonics in the vicinity of SPR of the C_{60}-containing plasma (40–70 nm) is estimated to be close to 10^{-4}. We demonstrate efficient broadband HHG in C_{60}-rich plasma plumes by introducing the techniques previously used for gas HHG. Further, we show that the phase modulation of fundamental radiation, both using the variation of the distance between the gratings in the compressor, and using the laser-plasma-induced self-phase modulation and spectral shift, can considerably change the spectral distribution of harmonics. All these changes in fundamental wave characteristics allow dramatic manipulation of the harmonic spectrum and intensity at well-defined conditions of fullerene plasma. We show, using simulations based on time-dependent density functional theory, the influence of collective excitations on enhancement of HHG near SPR. Further, simulations of resonant HHG by means of a multiconfigurational time-dependent Hartree–Fock approach for three-dimensional fullerene-like systems are shown.

5.1. First Observation of HHG in Fullerene Plasma

To study high-order nonlinearities through HHG, fullerene-containing targets were placed inside a vacuum chamber to ignite the plume by laser ablation of these structures. The targets used in these studies were (i) a C_{60} powder glued onto glass slides, (ii) 1 mm thick films of a C_{60} suspension in polymethylmethacrylate (PMMA) at different concentrations of the fullerenes, and (iii) thin films of deposited C_{60}. Other targets, which were used for comparison with C_{60}-containing plumes, were bulk carbon, soot powder, bulk indium, and bulk silver. The latter two targets have previously proved to be the most efficient media for production of plasma plumes, in which the highest HHG efficiency was observed. In one fullerene sample, C_{60} powder was mixed with epoxy and fixed onto glass substrates leading to an inhomogeneous distribution of fullerene clusters. Another sample was a fullerene film on a glass substrate. The film was grown by evaporating C_{60} powder in a resistively heated oven at 600 °C. The effused beam of C_{60} molecules was deposited onto a glass substrate maintained at liquid nitrogen temperature. The growth conditions, which depend on the rate of evaporation and deposition time, determined the thickness and the quality of the film. The thickness of the film used in the experiment was a few micrometers.

Figure 5.1 shows the harmonic spectra obtained using 30 fs pulses for ablation of a bulk carbon target, C_{60} powder fixed in epoxy on silver, and a C_{60} film. HHG produced in the ablation plume of bulk carbon targets exhibits a plateau-like harmonic spectrum up to the 25th order. To understand the origin

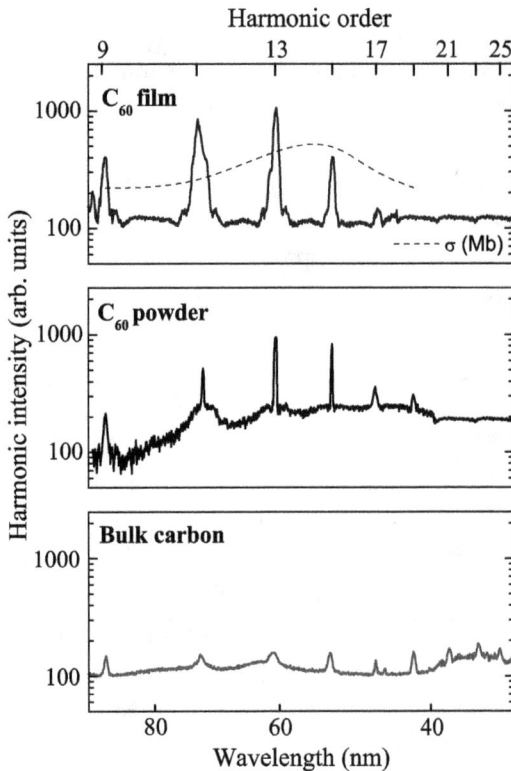

Fig. 5.1 Harmonic spectra obtained in the plasma plumes produced from a bulk carbon target, C_{60} powder-rich epoxy, and a C_{60} film. The dashed curve in the top panel corresponds to the photoionization cross sections near plasmon resonance. Adapted from [3] with permission from American Physical Society.

of HHG, the structure of the deposited carbon debris was studied. The absence of nanoparticles in the ablation plume of the bulk carbon target suggests carbon monomers as the source of harmonics. No specific enhancement in the harmonic yield or extension of cutoff was observed in the case of the bulk carbon target [3].

The harmonic spectra from targets containing C_{60} powder in epoxy and C_{60} film are significantly different in comparison with the bulk carbon target under identical experimental conditions. (i) Harmonics lying in the spectral range of SPR in C_{60} (20 eV, $\lambda = 62$ nm) are enhanced. (ii) The harmonic yields are larger by a factor of 20–25 for the 13th harmonic. (iii) The harmonic cutoff in C_{60} is lower (19th order) than in carbon but extends beyond the value (11th order) predicted by the three-step model. (iv) The 11th and 13th order harmonics in C_{60} are more intense than the 9th harmonic. Though the sensitivity of our detection system decreases for longer wavelengths at around 70 nm, in most cases, where various bulk targets and atoms were used, we observed a considerably stronger 9th harmonic compared with higher-order harmonics.

Increasing the intensity of the femtosecond driving pulse did not lead to an extension of the cutoff in fullerenes, which is a sign of saturation of the HHG in this medium. Moreover, at relatively high laser intensities, a decrease in harmonic output was observed. At such intensities, the multiple ionization of C_{60} leads to high free-electron density causing phase mismatch. Similarly, an optimal probe pulse intensity exists above which harmonics in C_{60} became weaker due to fragmentation of fullerenes, increase of the free electron concentration, phase mismatch, and self-defocusing. For fullerene targets, a heating pulse intensity of 2×10^9 W cm^{-2} was used, ten times lower than for the bulk carbon target. By calibrating our detection system using a technique previously reported [15], the efficiency of the 11th to 15th harmonics (between 50 and 70 nm) from the fullerene plume was estimated to be $\sim 10^{-5}$.

We now address the source of high-order harmonics in fullerene targets. The spatial characteristics of the targets were analyzed prior to laser ablation and compared with the ablated material debris deposited on nearby substrates (glass, aluminum foil, or silicon wafer). Surface morphology of the C_{60} film and the powder in epoxy was analyzed using an atomic force microscope. The structure of the C_{60} film is close to the crystalline shape with mean sizes of crystallites in the range of 80–200 nm. The sample containing C_{60} powder in epoxy suggests that C_{60} aggregates have sizes in the range of 200–600 nm. In comparison, the size of a single C_{60} molecule is less than 1 nm. The morphology of the debris due to ablation of the C_{60} film and powder were imaged using a scanning electron microscope and an atomic force microscope, respectively. The ablation debris contains the same aggregated particles as those prior to ablation. It was therefore concluded that fullerene clusters are responsible for HHG. No harmonics were observed in experiments during ablation of pure epoxy and the substrates alone without fullerenes.

The structural integrity of the fullerenes ablated off the surface should be intact until the probe pulse arrives. So, the heating pulse intensity is a very important parameter and was kept between 2×10^9 and 8×10^9 W cm^{-2} in the experiments. At lower intensities, the concentration of clusters in the ablation plume is low, while at higher intensities one can expect fragmentation. The temperature at the surface after the absorption of a 1 mJ heating pulse was estimated to be in the range of 600–700 °C, which was above the evaporation threshold of fullerenes ($\sim$300 °C) but below the temperature of fragmentation ($\sim$1000 °C). This estimation is valid for both types of fullerene targets. More details on the morphology of ablated and deposited fullerenes will be presented below.

Fullerene films produced more intense and stable harmonics with low shot-to-shot variation compared to the powder–epoxy mixture. This is due to the homogeneous distribution of particles in the film. In both types of fullerene targets the density of the ablation plume decreases for successive laser shots due to evaporation of C_{60} from the ablation area. As a result, the harmonic intensity decreases, as shown in Fig. 5.2 for C_{60} film. After about ten shots at the same target position, harmonic generation almost disappeared, unless the sample was moved to a fresh spot on the fullerene-containing target. To maintain the stability of the HHG process from C_{60} film, the film was moved after a few shots to avoid reduction of fullerene concentration in the ablation plume.

Fullerene density in the interaction region is a critical parameter in efficient generation of high-order harmonics. At high densities, phase mismatch and absorption of harmonics begin to dominate and shape the harmonic spectrum. Exact measurement or calculation of fullerene concentration in the ablation plume is difficult. In bulk targets, simulation techniques based on the hydrodynamic code HYADES can accurately predict the atomic and ionic concentration [16]. However, when extended to nanoparticle-rich targets,

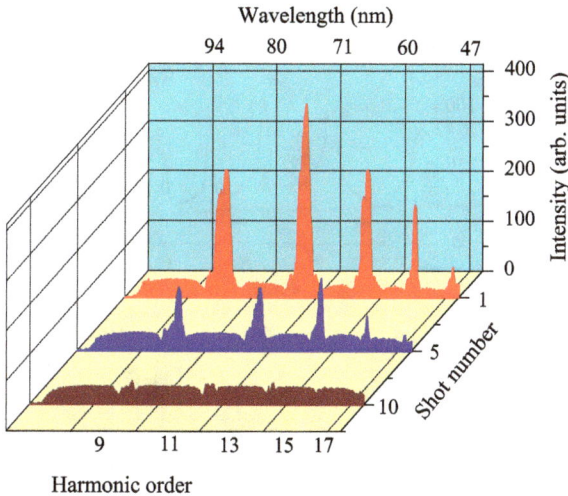

Fig. 5.2   Variation of harmonic spectra observed at consecutive shots on the same spot on a fullerene film. Reprinted from [3] with permission from American Physical Society.

simulations provide only a rough estimate of the density due to lack of information on the absorbance of these materials. For C_{60} film, the fullerene density was estimated to be no less than $5 \times 10^{16} \, \mathrm{cm}^{-3}$.

5.2. Influence of Various Experimental Parameters on HHG Efficiency in Fullerene Plasma

Figure 5.3 shows harmonic spectra in C_{60} for different delays between the ablation pulse and the 30 fs probe pulse. The HHG by ablation of bulk materials is greatly influenced by the temporal delay between the heating and probe pulses, as it alters the plasma density and length in the interaction region. To study its influence, we varied the delay from 18 ns to approximately 100 ns. The measurements showed no significant changes in the harmonic intensities in C_{60} for delays of 22 ns and 63 ns (see Figs. 5.3a and 5.3b). Also shown here is the harmonic spectrum from chromium plasma obtained at 63 ns delay (Fig. 5.3c). Overall, the two delays produce approximately equal harmonic intensities, with a two-fold increase of harmonic efficiency for the shorter delay [12].

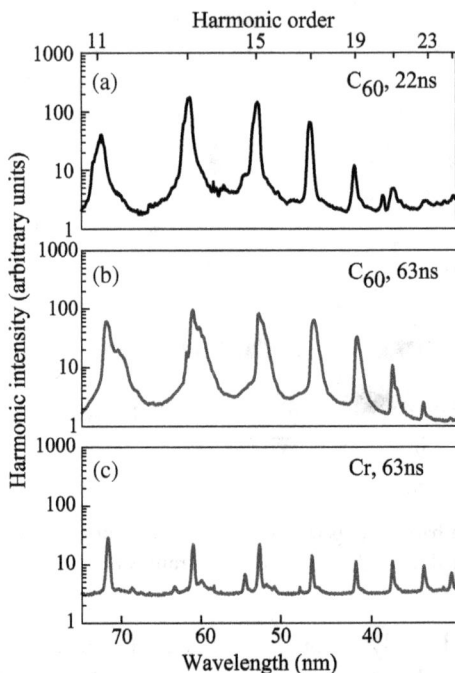

Fig. 5.3 Harmonic generation observed in C_{60} plasma at (a) 22 ns and (b) 63 ns delays between the heating and probe pulses and (c) in chromium plasma. Reprinted from [12] with permission from American Physical Society.

Note that for bulk targets, such as carbon, chromium and manganese, no harmonics were observed from their plasmas when very short delays ($\sim$6 ns) were used, in contrast to the case of C_{60}. This can be attributed to the nonoptimal plasma conditions in the case of the bulk target, since it requires time for the plasma to ablate on the bulk surface and expand into the area where the femtosecond beam interacts with the plasma. This can also be inferred from the lower heating picosecond pulse intensity ($I_{pp} \approx 2 \times 10^9$ W cm^{-2}) used for HHG from the C_{60}-rich target, compared with that used for bulk targets ($I_{pp} \approx 10^{10}$ W cm^{-2}).

One can postulate that short delays lead to more favorable evaporation conditions and higher particle density for the cluster-rich medium compared with the monatomic medium, thus resulting in a higher harmonic yield. In most cases of heavy bulk targets, the strong harmonics were observed using longer delays (40–70 ns). The use of light targets (boron, beryllium, lithium) showed an opposite tendency, where one can obtain effective HHG for shorter delays. The optimization is related to the presence of an appropriate amount of particles at the area of femtosecond pulse focusing, which depends on the propagation velocity of the plasma front. For C_{60}, one can expect the optimization of HHG at longer delays due to the larger weight of the fullerene particles. However, one has to admit the possibility of the presence of fragments of C_{60} in the plume, in which case the density of the medium in the area of interaction with the laser pulse becomes sufficient, even for shorter delays.

An interesting feature of the fullerene harmonic spectra is that the spectral width is about three to four times broader than for those generated in monoatom-rich plasmas (1.2 nm and 0.3 nm full width at half maximum, respectively). For comparison, Fig. 5.3c shows the harmonic spectra from chromium ablation. The broader width of the harmonics can be explained by self-phase modulation and chirping of the fundamental radiation propagating through the fullerene plasma. Broadening of the probe pulse bandwidth causes broadening of the harmonic bandwidth. The variation of harmonic bandwidth with delay can be explained by the higher density of the fullerene plasma for the longer delay and consequently stronger self-phase modulation of the fundamental radiation followed by broader width of harmonics.

At a relatively strong heating pulse intensity for fullerene film ($I_{pp} >$ $1 \times 10^{10}\,W\,cm^{-2}$), only the plasma spectrum was observed, without any sign of harmonics, as shown in Fig. 5.4a. The XUV spectra emitted by the plasma created in a vapor of C_{60} molecules have been studied in [17]. Although the experimental conditions are different, the plasma emission spectra are comparable to that shown in Fig. 5.4a. The spectra show multiple lines in the range of 18–26 nm associated with ionized fragments of C_{60} (in particular from C^{3+}–C^{5+}) together with the ionic lines near 38 and 54 nm. In contrast to the fullerene plasma spectrum, the carbon plasma spectrum at the same conditions of excitation showed ionic lines only from lower-charged ions, as shown in Fig. 5.4b. The origin of this difference was attributed to multi-electron dissociative ionization of molecules as a complex dynamic sequence of events. C_{60} has demonstrated both direct and delayed ionization and fragmentation processes and is known to survive even in intense laser fields, which can be attributed to its large number of internal degrees of freedom, which leads to the fast diffusion of excitation energy [6, 18]. At 796 nm, multiphoton

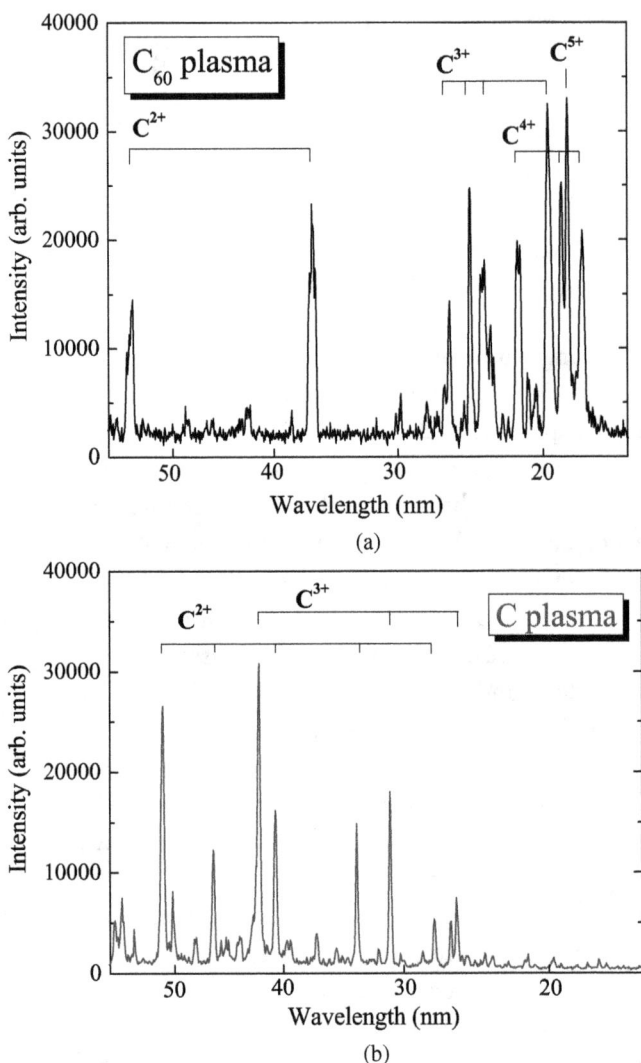

Fig. 5.4 Plasma spectra of laser-ablated (a) C_{60} film and (b) carbon bulk target observed at high heating pulse intensity ($I_{pp} = 2 \times 10^{10}$ W cm^{-2}). Reprinted from [12] with permission from American Physical Society.

ionization is the dominant mechanism leading to the ionization of C_{60} in a strong laser field. The collective motion of the π electrons of C_{60} can be excited by a multiphoton process. Since the laser frequency is much smaller than the resonance frequency of π electrons, barrier suppression and multiphoton ionization are the dominant mechanisms leading to ionization in a strong laser field.

Another important parameter that affects the stability of the HHG process is the thickness of the fullerene target. Relatively stable harmonic generation with low shot-to-shot variation in harmonic intensity was observed by moving the fullerene film deposited on the glass substrate after several laser shots. This avoids a decrease in the fullerene density due to ablation of the thin film. The number of laser shots at the same target position that resulted in stable harmonic emission decreased drastically with the film thickness. For example, in a 10 μm film, the harmonic emission disappeared after 70–90 shots, whereas in a 2 μm film the number of laser shots is reduced to 5–7.

The dependence of the harmonic emission from C_{60} on the polarization of the probe pulse was investigated. This also enabled us to differentiate the plasma emission from the HHG process. The HHG is highly sensitive to laser polarization, since the trajectories of the recolliding electrons are altered significantly, thereby inhibiting the recombination process. It was observed that the harmonic signal drops rapidly and disappears with ellipticity of the laser polarization. Figure 5.5 shows the HHG spectra in the cases of linear

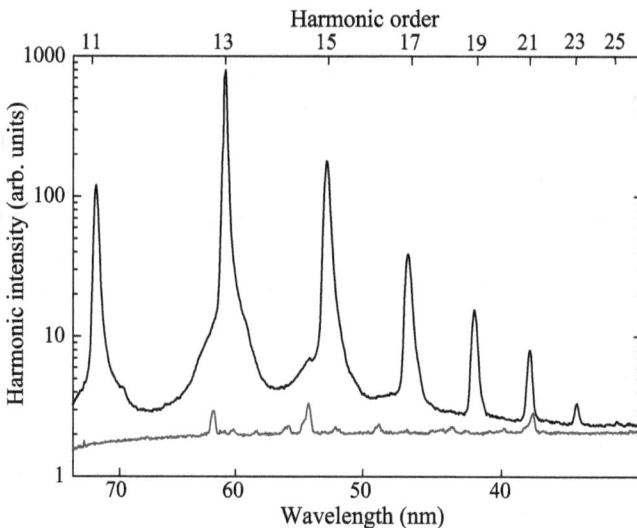

Fig. 5.5 Harmonic spectra obtained in C_{60}-rich plasma, for linearly (upper curve) and circularly (bottom curve) polarized driving laser. Reprinted from [12] with permission from American Physical Society.

and circular polarizations of the driving laser. For circular polarization, as expected, the harmonic emission disappears and the resulting background spectrum corresponds to the plasma emission.

Does the influence of plasmon resonance on HHG in fullerene plasma depend on the wavelength of the driving field? To address this question, HHG using the second harmonic (396 nm, 4 mJ, 35 fs) of the probe pulse (793 nm, 30 mJ) was studied. In this case, a 2 mm thick KDP crystal was inserted in front of the focusing lens, thus excluding the exact temporal and spatial overlap of the fundamental and SH pulses in the plasma area. The relatively low SH conversion efficiency did not allow us to achieve the laser intensities attained with the 793 nm fundamental laser. As a result, harmonics only up to the ninth order of the 396 nm probe pulse were obtained, while simultaneously generating harmonics using the 793 nm laser. Harmonic generation using two driving pulses (793 nm and 396 nm) did not interfere with each other, since the two HHG processes occurred in different regions of the laser plasma. Quite another pattern of harmonic spectra appeared when the driving radiation was converted in a thin (1 mm) crystal placed after the focusing lens. The details of these experiments will be discussed in Section 5.4.

Figure 5.6a shows the HHG spectrum from C_{60} fullerene optimized for the SH probe pulse. The energy of the SH is $\sim 1/7$th of the fundamental. One can see the enhancement of the seventh harmonic (which is within the range of the SPR of C_{60}) compared with the fifth harmonic. This behavior is similar to that observed for the 793 nm probe pulse. For comparison, in Fig. 5.6b the optimized harmonics generated using the 396 nm radiation and the weak harmonics from the 793 nm radiation in manganese plasma are presented. One can see a decrease in harmonic intensity from the manganese plume for each subsequent order, which is a common case when one uses a nonlinear optical medium containing atomic or ionic particles. These studies confirmed that, independently of the probe pulse wavelength, the harmonics near SPR in C_{60} are always enhanced.

For 48 fs driving pulses, when the fullerene-rich target ablated and generated harmonics from the plasma, the cutoff was extended up to the 29th order ($\lambda = 27.6$ nm). The intensity of the harmonics from C_{60}-rich plumes was considerably stronger than those generated from plasma rich with single particles, created on the surface of bulk targets under the same experimental conditions. For comparison, soot powder and bulk graphite were used. Other

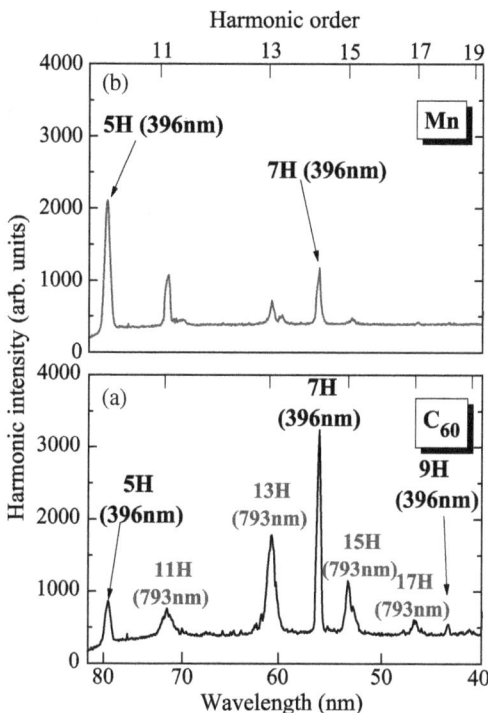

Fig. 5.6 Harmonic spectra from (a) C_{60} and (b) manganese plasma, when both the 793 nm and 396 nm pulses were simultaneously focused on the laser-produced plasma. Reprinted from [12] with permission from American Physical Society.

bulk materials used were indium and silver. Indium plasma has been proven as the medium where the strongest resonance-induced harmonic (13th order, 61.5 nm) is generated with conversion efficiency close to 10^{-4} [19].

This harmonic radiation was comparable with the harmonics from C_{60}-containing plasmas (Fig. 5.7). From another angle, the harmonics generated in a silver plasma plume have shown the strongest yield among other ablated targets in the range of the 30th to 50th orders ($\sim$25–15 nm) [20]. Harmonics from the plasma produced on bulk carbon and powder soot targets were also compared with fullerene harmonics to distinguish the peculiarities and advantages of the fullerene-containing nonlinear medium. In all of these comparative studies, the C_{60}-rich plasma demonstrated better HHG efficiency.

Placing an aperture in front of the focusing lens led to a significant change in the harmonic intensity, while the pulse energy before the aperture was

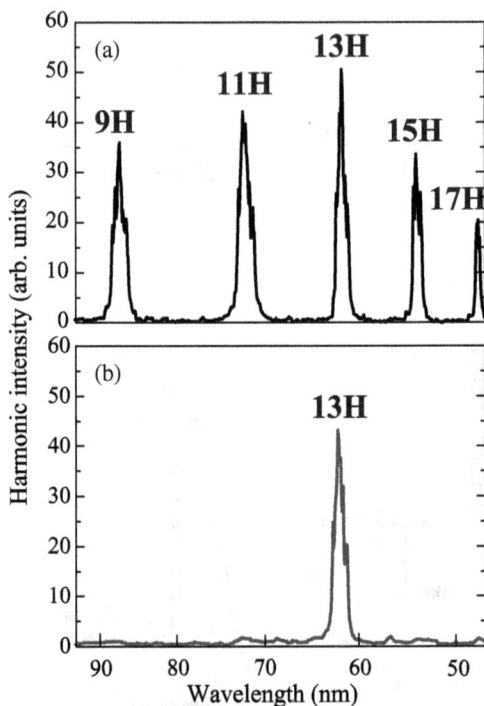

Fig. 5.7 Comparison of harmonic intensity in the cases of (a) fullerene plasma and (b) indium plasma. One can see the equality between the resonance-induced enhancement of the 13th harmonic generated in indium plasma and the group of harmonics from fullerene plasma in the range of SPR of C_{60}. Reprinted from [13] with permission from American Institute of Physics.

kept the same. A two-fold decrease in beam size led to both a four-fold decrease in pulse energy (assuming it to be uniform across the beam) and an increase in beam divergence. The latter led to intensity increase of the 11th and 13th harmonics compared to the aperture-less configuration, even though the energy of the pump laser was decreased by a factor of four. Such an increase of specific harmonics was also observed in various targets containing fullerenes. This change in the harmonic yield can be attributed to better phase matching conditions for some groups of harmonics.

As mentioned earlier, the 13th harmonic from indium plasma considerably exceeds other neighboring harmonics due to closeness with resonance transitions possessing high oscillation strength. A comparable HHG efficiency for the harmonics in the SPR range of C_{60} and the 13th harmonic from the plume created on the surface of an indium target has been achieved. Although no quantitative measurement of conversion efficiency in fullerene plasma was carried out in these experiments, the results presented in Fig. 5.7 show that the HHG efficiency at $\lambda \approx$ 50–90 nm was in the range of 10^{-4}, analogous to the conversion efficiency of the single harmonic in indium plasma.

The fullerene powder directly glued onto a glass surface could survive for a longer time during interaction with heating pulse radiation than the fullerene–PMMA films. In particular, analysis of HHG for the powder showed that harmonics could be observed during approximately 90 shots on the same spot. Less stability was observed in the case of the film containing fullerenes in PMMA. The harmonics from this target survived only for 15–25 shots. The difference in the stability of HHG was mostly due to the different thickness of fullerenes in the cases of the C_{60} powder glued on the substrate and C_{60}–PMMA film. The appearance of craters on the fullerene-containing targets also led to a change of optimal conditions for HHG in laser plasma.

Most of the experiments were carried out when femtosecond radiation was focused before the plasma plume. However, the HHG efficiency in fullerene plasma was approximately two times stronger in the case of the focusing of laser radiation after the plasma plume compared with the focusing before the plasma plume.

Figure 5.8a shows high-order harmonic spectra obtained from a fullerene plasma using 150 fs laser pulses at appropriate excitation of the fullerene powder-containing target. Over-excitation of the target led to generation of strong plasma emission from the over-ionized fullerenes and their disintegrated parts, together with considerable decrease of harmonic emission, when only few weak harmonics can be identified in the observed spectra (Fig. 5.8b). The mechanism that leads to the deterioration of HHG is related to the appearance of considerable numbers of free electrons, leading to the phase mismatch of harmonics.

In these studies using relatively long (150 fs) laser pulses, the harmonic cutoff from C_{60} plasma was extended up to the 33rd order, compared to that at 19th order (for 30 fs pulses) and 29th order (for 48 fs pulses). This extension was achieved by optimizing the experimental conditions for producing the fullerene plasma and laser–matter interaction. For a fullerene plume, increasing the intensity of the 150 fs probe pulse did not help extension of the cutoff above the 33rd harmonic, which is a sign of HHG saturation in this medium.

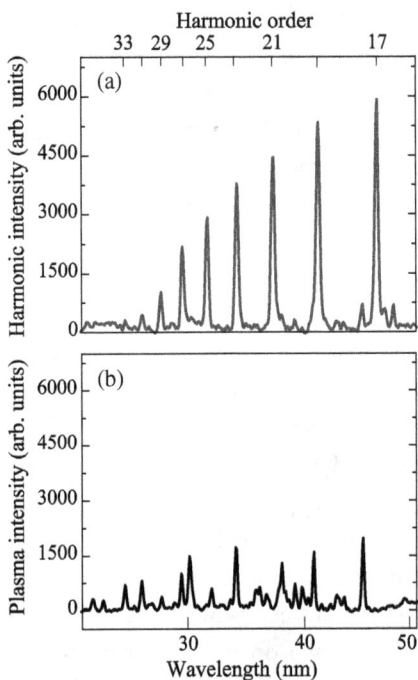

Fig. 5.8 (a) Harmonic spectrum obtained from fullerene plasma at optimal excitation of the target. (b) Plasma spectrum obtained at over-excitation of a fullerene-containing target. Reprinted from [14] with permission from Springer Science+Business Media.

Moreover, at relatively high intensities of the femtosecond laser, a decrease in harmonic output was observed, due to the prevalence of restricting factors (such as over-ionization, higher free-electron density, self-defocusing, and phase mismatch). A similar phenomenon was observed when the heating pump pulse intensity on the surface of C_{60} targets was increased above some optimal value (Fig. 5.8b).

5.3. Studies of Harmonic Modulation from Fullerene-Rich Plasmas

Modulation of harmonic spectra was analyzed by (i) introducing a chirp by changing the distance between the gratings in the compressor, (ii) phase modulation of the probe radiation by introducing a 10 mm thick BK-7 glass plate between the focusing lens and the plasma plume, and (iii) phase modulation during propagation of intense laser radiation through the fullerene-containing plasma. In all of these cases, a broadening of harmonic bandwidth and a shift of the central wavelength toward the blue or red sides of the spectrum was observed. A study of the optimized HHG from the C_{60} plume by changing the chirp of the driving radiation, without modifying the probe laser spectrum, is presented below. The chirp of the driving laser pulse was varied by adjusting the separation of the gratings in the pulse compressor. At the chirp-free condition, the laser pulse duration was measured to be 48 fs. The introduction of positive or negative chirp led to an increase in the pulse duration up to 300 fs.

During HHG in C_{60} plasma, the harmonics shifted toward longer wavelengths in the case of the positively chirped probe pulses, when the leading edge of the pulse had a red component compared with the trailing edge. This effect can be explained by the wavelength change on the leading edge of the chirped laser pulse. The initial lower-intensity portion of the pulse created harmonics. As the pulse intensity reached its peak, the conditions for HHG in C_{60} plasma were spoiled. Thus, it is the leading edge of the pulse that contributes to HHG. The harmonics produced with positively chirped laser pulses were redshifted because the harmonics produced in the leading edge of the laser pulse come from the red part of the laser spectrum. The same can be said about the blueshifted harmonics produced by the negatively chirped pulses.

No significant influence of self-phase modulation (SPM) on the spectral distribution of harmonics was expected as the experimental conditions (low-density plasma, moderate laser intensities) restricted the possibility of the influence of a strongly ionized medium on the phase characteristics of the generated harmonics. A highly ionized medium, with electron density higher in the center than in the outer region, acts as a negative lens, leading to a defocusing of the laser beam in a plasma and hence to a reduction in the effective harmonic generation volume. In addition, the rapidly ionizing high-density medium modifies the temporal structure of the femtosecond laser pulse due to SPM. By keeping the laser intensity in the vicinity of plume close to the barrier suppression intensity of singly charged C_{60}, the condition was established when no significant ionization of the plasma by the driving laser pulse takes place.

In the case of SPM in a laser plasma, one can expect a considerable variation of the harmonic spectrum compared to the case of moderate intensities of

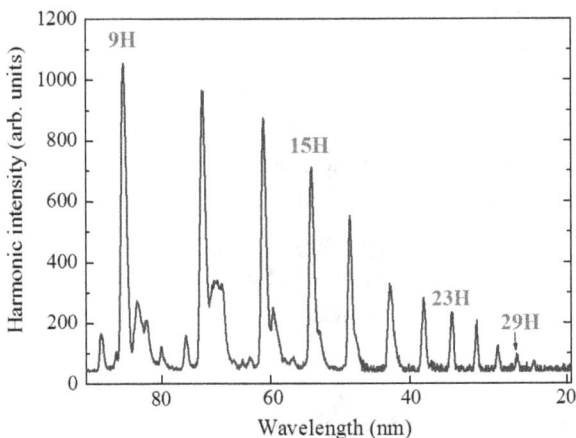

Fig. 5.9 Harmonic spectrum from a fullerene plasma plume in the vicinity of the SPRs of C_{60} and C_{60}^+ (55–65 nm) and beyond. Reprinted from [13] with permission from American Institute of Physics.

fundamental radiation, when no changes in either probe or harmonic spectra are expected. Such a variation of the harmonic spectra was mostly defined by the modulation of the probe pulse spectrum. A strong extension of harmonic spectral distribution toward the blue side (as seen in Fig. 5.9) was observed. It may be noted that these variations were emphasized in the vicinity of the broad SPRs of C_{60} and C_{60}^+ with the central wavelengths near 60 and 50 nm (i.e., close to the 13th and 15th harmonics of the 800 nm pump) and broad wings overlapping the range of the 9th and partially the 17th harmonics. Each of these harmonics possesses strong lobes on the short-wavelength side. Weak lobes were observed also for the harmonics beyond the SPR of C_{60}. In particular, the higher harmonics (i.e., above the 21st order) showed a smooth spectral distribution, which repeats the fundamental spectrum. The lobes beyond the SPR of C_{60} (in particular for the 17th and 19th harmonics) can be attributed to the broad SPR of C_{60}^+ ions ($\lambda = 50$ nm). These observations show that, in the vicinity of strong collective electron transitions of a nonlinear medium, one can expect considerable SPM and corresponding modulation of the harmonic spectra. It may be noted that some weak lobes were observed in the cases of graphite and soot plasmas as well. Probably, the plasma conditions for these media were sufficient to achieve SPM of the fundamental radiation.

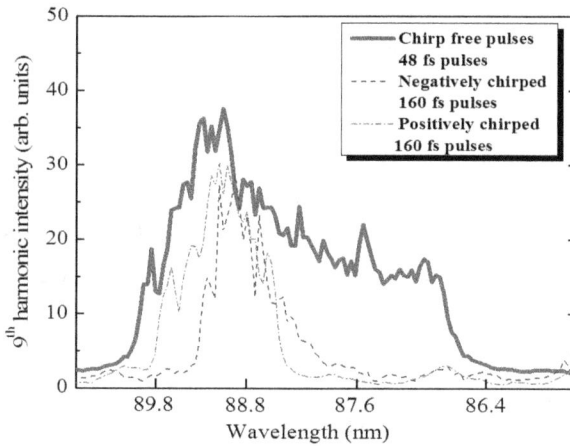

Fig. 5.10 Spectrum of the 9th harmonic from a fullerene plasma in the cases of chirp-free 48 fs pulses (solid line), positively chirped 160 fs pulses (dash-dot line), and negatively chirped 160 fs pulses (dashed line). Reprinted from [13] with permission from American Institute of Physics.

Analysis of the variations of these spectra at different chirps of the probe radiation showed a suppression of blue-sided lobes for both negatively and positively chirped pulses (see Fig. 5.10, dashed and dash-dotted lines). This suppression was caused by a decrease of probe pulse intensity of the chirped pulse leading to a corresponding decrease of the influence of SPM on the spectral distribution of the laser radiation propagating through the C_{60} plasma. In the absence of SPM, these harmonics show only a blueshift or a redshift depending on the chirp of laser pulse.

The variation of harmonic spectra in the case of the SPM induced by introduction of a 10 mm thick glass slab in the path of fundamental radiation between the focusing lens and the plasma plume was also studied. Analysis of the fundamental radiation spectrum showed a considerable change in the spectral distribution of laser radiation when the intensity of the laser radiation inside the glass slab reached 2×10^{11} W cm^{-2}. Appearance of a broadened blue-sided lobe in the laser spectrum was clearly observed in these studies. Further movement of the glass slab toward the focus area caused the enhancement of the blue-sided part of laser spectrum compared with the case of the initial spectrum. Note that no white-light generation was observed at these conditions.

All these variations of the laser spectrum were transferred to the harmonic spectra. The extension of harmonic spectra toward the shorter-wavelength side was observed. At the same time, a redshift of harmonics was also observed. This can be explained in terms of the longer wavelength generation in the rising part of the laser pulse due to SPM, which also gives a positive chirp, in addition to the positive chirp due to group velocity dispersion in the glass slab. So, the presence of red frequencies in the initial part of the laser pulse gives rise to the redshift (as was seen for positively chirped pulses without the glass slab).

In the case of the bulk carbon target, the harmonic lobes were either absent or considerably suppressed. At the same time, in the case of the soot target, the blue-sided lobes in the harmonic spectra were observed, which is probably due to the presence of some fullerenes expected in the soot.

5.4. Two-Color Pump for Harmonic Generation in C_{60}

Experiments and theoretical studies on two-color HHG in gas jets and cells have previously been carried out for frequency ratios 2:1 [21] of the two fields, with the strengths of the two fields either widely different or comparable. The reported experimental results have shown that the presence of the second field strongly modifies the harmonic spectrum. It may be noted that, in some of gas HHG experiments, the use of two-color probe pulses led to both the enhancement of harmonic efficiency and the extension of harmonic cutoff. In experiments with fullerene plasma plumes, no extension of harmonic cutoff was observed; the cutoff remained the same for both pump schemes (i.e., single- and two-color), and mostly depended on the conditions of excitation of the C_{60}-rich target. The maximum observed harmonics in both the schemes were in the range of the 29th harmonic using 48 fs driving pulses.

Insertion of a 1 mm thick SH crystal (KDP) in the beam path after the focusing lens led to generation of enhanced harmonic yield and the appearance of even and odd harmonics with approximately equal intensities (for lower orders of harmonics, see Fig. 5.11) [22]. Introduction of a UV filter after the SH crystal led to generation of a few odd harmonics (5th and 7th: Fig. 5.11a) from the SH field (i.e., generation of the 10th and 14th harmonics of the probe radiation), while the intensities of these harmonics were weaker than those in

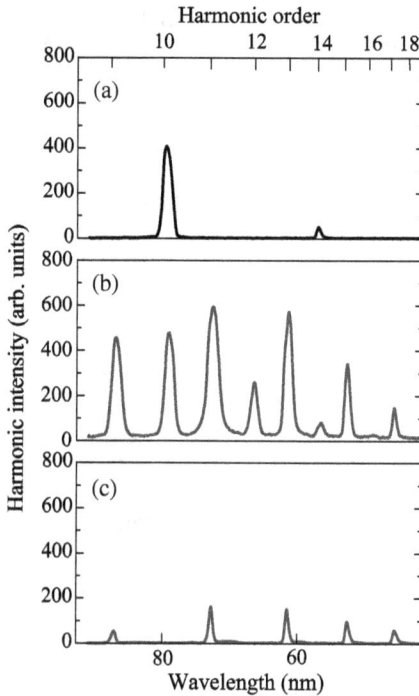

Fig. 5.11 Harmonic spectra generated in a C_{60} plasma plume in the case of (a) single-color SH pump (400 nm), (b) two-color pump (800 nm + 400 nm), and (c) single-color fundamental pump (800 nm). Reprinted from [22] with permission from Springer Science+Business Media.

the case of the two-color pump (Fig. 5.11b). A considerable enhancement of HHG efficiency was observed in the two-color case compared to the single-color 800 nm pump (Figs. 5.11b and 5.11c). The enhancement factor in that case was in the range of four and eight, depending on the harmonic order. Even harmonics of the fundamental radiation up to 16th order were obtained in these studies. The fact that only the 5th and 7th harmonics were observed (Fig. 5.11a) in the experiments with the single SH pump can be attributed to the small conversion efficiency in the SH wave (2%).

Analogous generation of odd and even harmonics was observed also in the case of bulk targets [23]. However, the efficiency of even harmonics was much lower than that in fullerene plasma plumes. Figure 5.12 shows the harmonic spectra from indium and C_{60} plasma plumes for the two-color pump. One can see the strong even harmonics from the fullerene plasma, while in the case

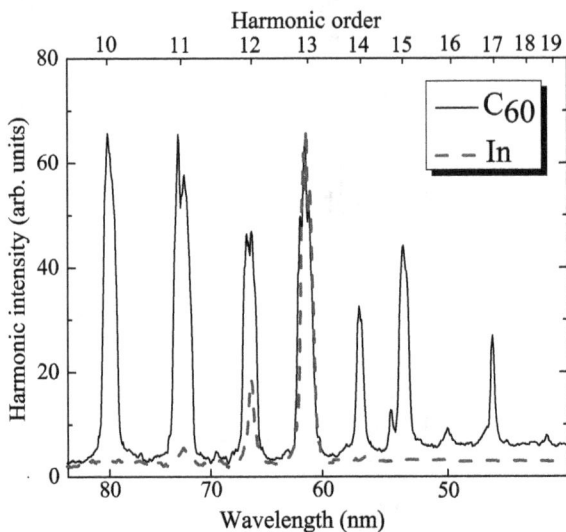

Fig. 5.12 Harmonic spectra for the two-color pump from C_{60} plasma (solid line) and indium plasma (dashed line). Reprinted from [22] with permission from Springer Science+Business Media.

of indium plasma, only one (i.e., the 12th) even harmonic is visible in the harmonic spectrum.

5.5. Analysis of the Morphology of Fullerene Targets and Ablated Materials

Ablation-induced nanocluster formation in laser plumes has been documented in several experiments [24–26]. In the case of bulk target ablation, care is taken to create conditions such that the laser energy is accumulated for a short period at a small area to maintain the conditions of nonequilibrium heating. In that case, the extremely heterogeneous conditions help in creating clusters in the small areas of heated samples. One can maintain the conditions so that the aggregated atoms do not disintegrate during evaporation from the surface [27]. The maintenance of the original properties of clusters allows one to analyze the optical and nonlinear optical properties of cluster-containing laser plasma at well-defined conditions. This becomes extremely important in the case of large-sized molecular targets, such as fullerenes. These clusters show strong resistance to disintegration under the action of laser radiation, due to dissipation of absorbed energy among the multiple transition bands of the C_{60} molecule. In this connection, careful analysis of the C_{60} plasma debris can give indirect confirmation of the presence of these large molecules in the plasma plume.

Below, experimental studies of the structural modifications of material during laser ablation of C_{60}-containing targets are described. The morphology of the initial material and ablated clusters is studied by analyzing the debris deposited on nearby substrates. This technique allowed the optimization of laser ablation parameters for maintaining the fullerenes in the laser plumes [13].

To create the ablation, only the heating beam was focused on a target placed in a vacuum chamber. The spot size of this beam on the target surface was maintained in the range of 0.5–0.8 mm. The laser energy density during ablation was kept at 0.4–1 J cm^{-2}. The chamber was maintained at a pressure of 8×10^{-6} mbar. The debris was deposited from the plasma plume onto a silicon substrate and copper grids with carbon films, placed nearby, which were then analyzed using transmission electron microscopy.

Targets containing fullerene clusters were ablated. The commercially available mixtures of fullerenes (98% of C_{60} and 2% of C_{70} powder) were glued onto glass substrates or mixed with PMMA. Blocks of C_{60} in polymer PMMA were made by mixing the two components in a solvent mixture (trichlorobenzene and toluene in 80:20 ratio by volume). The ratio of C_{60}:PMMA was kept at ~50:50 by weight. After dissolving the two components in the solution, the suspension was poured onto glass slides and dried in a vacuum oven at 70°C. By this method, a homogeneous C_{60}–polymer composite film of thickness about 1 mm was obtained.

Initially, the structure of the C_{60} powder used in these studies was analyzed with a high-resolution transmission electron microscope (HRTEM). The C_{60} powder consisted of aggregates of fullerene clusters in the shape of crystallites. Sample preparation for HRTEM observation was performed by preparing a suspension of the sample in methanol that was then dropped onto a copper-supported carbon film. To reduce agglomeration, the dilute fullerene emulsion was subjected to ultrasonic dispersion for approximately 10 minutes. The transmission electron microscope (Philips CM200) was operated at 200 kV accelerating voltage for microstructural study. The sizes of the aggregated C_{60} particles covered a range from 30 to 700 nm. It is likely that particles with

Fig. 5.13 (a) HRTEM image of a C_{60} powder agglomerate before deposition. (b) HRTEM image of deposited debris of C_{60} after strong excitation ($I_{pp} = 5 \times 10^{10}\ \mathrm{W\,cm^{-2}}$) of fullerene-containing target. The scale bars on the images correspond to 2 nm. In the insets, the Fourier transform patterns of the C_{60} crystalline nanopowder and debris are shown. Reprinted from [13] with permission from American Institute of Physics.

similar or close crystallographic orientations form bulky crystals or quasi-crystals with modulated surfaces and regular shapes.

Figure 5.13a shows an HRTEM image of the edge of a C_{60} aggregated cluster powder. At some places, the regular spacing of the lattice planes was observed to be 0.6 and 0.8 nm for the C_{60} clusters, which is consistent with the lattice spacing of these face-centered cubic structures [28]. The crystalline state of the fullerene particles was verified by a Fourier transform performed on the HRTEM image (see the inset in Fig. 5.13a). Due to the random position of the C_{60} powder in the grid, different electron diffraction patterns were obtained from the same material. Analogous features remained in the case of the HRTEM of deposited debris of the fullerene powder after laser ablation at moderate intensities ($\leq 7 \times 10^{9}\ \mathrm{W\,cm^{-2}}$).

Another pattern in HRTEM of the debris of ablated fullerene powder appeared at heating pulse intensities above $1 \times 10^{10}\ \mathrm{W\,cm^{-2}}$. In that case, a different spacing of the lattice planes created on the surface of substrates and grids after laser ablation was observed (Fig. 5.13b). The regular spacing of these lattice planes was 0.36 nm, which is consistent with the inter-planar lattice spacing for graphitic layers (0.34 nm). The HRTEM image presented in Fig. 5.13b indicates microstructure typical of carbon black: an intermediate structure between amorphous and fully graphitized carbon. The corresponding Fourier transform pattern also revealed a drastic difference with regard to the above-shown crystalline structure of the studied C_{60} samples. This pattern (see the inset in Fig. 5.13b) is a characteristic of the presence of an amorphous graphite structure. These results are consistent with previously reported analysis of the morphology of fullerene and graphite aggregates [28]. These studies also revealed the range of heating pulse intensities, which could be useful for maintaining the fullerenes in the laser plumes after laser ablation of C_{60}-containing targets [13].

5.6. Theoretical Calculations of HHG in Fullerenes

Theoretical studies on HHG from C_{60} involved extending the three-step model [29] by analyzing an electron constrained over the surface of a rigid sphere, with geometrical parameters similar to those of the C_{60} fullerene [30], and using dynamical simulations [31]. In the latter case, higher-order harmonics were shown to be due to multiple excitations and could be easily generated even with a weak laser field. Both studies reveal how HHG can be used to probe the electronic and molecular structure of C_{60}. At the same time, theoretical investigation of such systems is hampered by the fact that the Hamiltonian of HHG is time dependent and the systems consist of many electrons.

The efficiency of the HHG process can be understood in terms of three length parameters. For optimum HHG, the length of the nonlinear medium L_{med} should be (i) larger than the coherence length $L_{coh} = \pi/\Delta k$, which is defined by the phase mismatch between the fundamental and harmonic fields ($\Delta k = k_q - qk_0$ where k_q and k_0 are the harmonic and fundamental wave vectors, respectively) and depends on density and ionization conditions, and (ii) smaller than the absorption length of the medium $L_{abs} = 1/\rho\sigma$, where ρ is the atomic density and σ is the ionization cross section. When propagating through a medium, the wave vector of light with vacuum wavelength λ is given by

$$k = \frac{2\pi n_g(\lambda)}{\lambda} \tag{5.1}$$

with the index of refraction, $n_g = 1 + P\,\delta(\lambda)$, with pressure, P, in atmospheres and a suitable gas dispersion function, $\delta(\lambda)$. In general, empirical relationships are difficult to deduce at wavelengths shorter than the ultraviolet, although calculated data can be incorporated.

For sufficiently large laser intensities, partial ionization of the gas medium occurs, resulting in a modified index of refraction:

$$n_g = 1 + (1 - \eta)P\delta(\lambda) + (1 - \eta)n_2 I - \eta P N_{atm} r_e \lambda^2 / 2\pi \qquad (5.2)$$

where N_{atm} is the number density at atmosphere pressure, η is the ionization fraction, and r_e is the classical radius of an electron. The nonlinear index of refraction, n_2, is sufficiently small at this intensity that it will not be considered further. One can assume an ionization fraction of $\eta = 0.5$ for both carbon and C_{60}.

The phase mismatch between the fundamental and the qth harmonic can then be written

$$\Delta k = \eta P N_{atm} r_e (q\lambda_0 - \lambda_q) - \frac{2\pi(1-\eta)P}{\lambda_q}[\delta(\lambda_0) - \delta(\lambda_q)] \qquad (5.3)$$

The number of photons in the qth harmonic per unit time and area emitted on-axis is proportional to

$$N_{out} \propto \rho^2 A_q^2 \frac{4\rho^2 L_{abs}^2}{1 + 4\pi^2(L_{abs}^2/L_{coh}^2)} \left[1 + \exp\left(-\frac{L_{med}}{L_{abs}}\right) \right.$$
$$\left. -2\cos\left(\frac{\pi L_{med}}{L_{coh}}\right) \exp\left(-\frac{L_{med}}{2L_{abs}}\right) \right] \qquad (5.4)$$

where A_q is the amplitude of the atomic response approximated to be $(1-\eta)I^3$, where I is the laser intensity [32].

Due to increased absorption in C_{60}, one can expect a dip in the harmonic spectrum for the 11th to 15th harmonics. The calculations also indicate that the same harmonics produced in carbon vapor are not absorbed by the nonlinear medium. With an assumed medium length of 1 mm, theoretical spectra are obtained by inserting the appropriate wavelength-dependent index of refraction and dispersion data into Eqs. 5.2 and 5.3. For carbon, the index of refraction data were taken from [33] while the photoabsorption cross-section data were taken from [34]. For C_{60}, the photoabsorption cross-section data were taken from [35]. As there is no access to reliable index of refraction data

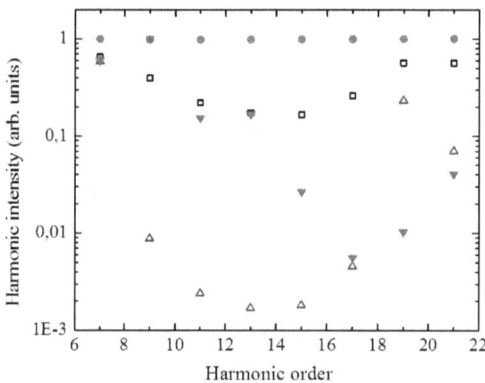

Fig. 5.14 Theoretical harmonic spectrum for C_{60}, considering absorption only (squares) and a combination of absorption and dispersion (open triangles). When dispersion is included, the predicted harmonic signal is quite small. The spectra for carbon are included for comparison. Identical results are obtained when using absorption only (circles) and absorption and dispersion (filled triangles). Reprinted from [12] with permission from American Physical Society.

for C_{60}, one can simply use a wavelength-dependent index of refraction that scales with the photoabsorption cross section.

Figure 5.14 shows the calculated harmonic spectra for C_{60} and carbon plasmas by considering only absorption (squares and circles respectively) and by including both absorption and dispersion (open and filled triangles respectively). In carbon vapor, the influence of absorption on the harmonic yield is negligible and as a result the overall harmonic spectrum is determined by dispersion. The harmonic yield decreases with increasing order as it becomes difficult to phase match higher orders.

In C_{60}, absorption of harmonics by the nonlinear medium is dominant due to large photoabsorption cross sections. The effect of dispersion seems to lower the HHG efficiency but does not affect the overall shape of the spectrum. As a result one expects the harmonic yield to decrease considerably near the SPR, if one does not consider the nonlinear optical influence of this resonance on the harmonic efficiency in this medium.

Photoionization cross sections of C_{60} are well known experimentally and theoretically. C_{60} exhibits a giant plasmon resonance at ~ 20 eV (around the 11th, 13th, and 15th harmonics of a Ti:sapphire laser). The absorption length was calculated using the estimated fullerene density in the interaction region and the photoionization cross sections (Fig. 5.15). It varies from 1 mm (for the

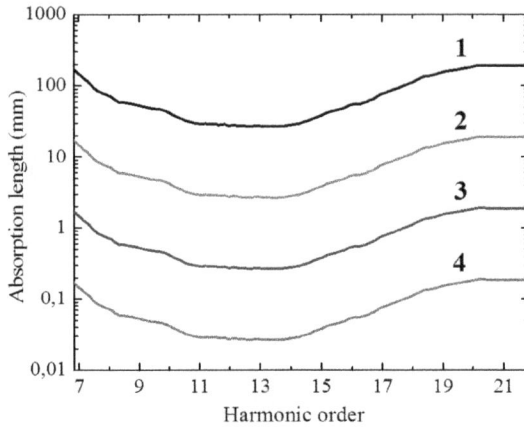

Fig. 5.15 Calculated absorption length of fullerene-containing plasma at different plasma concentrations as a function of harmonic order. (1) $N = 1 \times 10^{14}$ cm^{-3}; (2) $N = 1 \times 10^{15}$ cm^{-3}; (3) $N = 1 \times 10^{16}$ cm^{-3}; (4) $N = 1 \times 10^{17}$ cm^{-3}. Adapted from [12] with permission from American Physical Society.

7th and 17th harmonics) to 0.5 mm (for the 11th, 13th, and 15th harmonics) indicating that harmonics in the region of the plasmon resonance should be absorbed in the medium (whose length is estimated to be $\sim$0.8 mm). As was mentioned, due to this absorption one can expect a dip in the harmonic spectrum for the 11th to 15th harmonics. On the other hand, enhancements of these harmonics were observed. This is a signature of multi-electron dynamics in a complex molecule such as C$_{60}$ and has no atomic analogue. The calculations also show that harmonics produced in the bulk carbon target are not absorbed by the nonlinear medium. As a result, no modulation was seen in the harmonic spectrum of carbon in the plateau region.

To understand the origin of enhancement of harmonic yield near SPR, the interaction of monatomic carbon and the fullerene C$_{60}$ molecule with a strong laser pulse was investigated by means of time-dependent density functional theory [36]. In the TDDFT approach, the many-body time-dependent wave function is replaced by the time-dependent density $n(r, t)$, which is a simple function of the three-dimensional vector r. $n(r, t)$ is obtained with the help of a fictitious system of noninteracting electrons by solving the time-dependent Kohn–Sham equations:

$$i\frac{\partial}{\partial t}\phi_i(r, t) = \left[-\frac{\nabla^2}{2} + v_{KS}(r, t)\right]\phi_i(r, t) \qquad (5.5)$$

These are one-particle equations, so it is possible to treat large systems such as fullerenes. The density of the interacting system is obtained from the time-dependent Kohn–Sham orbitals:

$$n(r, t) = \sum_{i}^{occ} |\varphi_i(r, t)|^2 \tag{5.6}$$

$$v_{KS}(r, t) = v_{ext}(r, t) + v_{Hartree}(r, t) + v_{xc}(r, t) \tag{5.7}$$

Here $v_{ext}(r, t)$ is the external potential (laser field), $v_{Hartree}(r, t)$ accounts for the classical electrostatic interaction between the electrons:

$$v_{Hartree}(r, t) = \int d^3r' \frac{n(r, t)}{|r - r'|} \tag{5.8}$$

The exchange and correlation within the so-called adiabatic local density approximation (ALDA) was treated assuming that the potential is the time-independent xc potential evaluated at the time-dependent density:

$$v_{xc}^{adiabatic}(r, t) = \tilde{v}_{xc}[n](r)|_{n=n(t)} \tag{5.9}$$

For all calculations the code OCTOPUS [37] was used with norm-conserving nonlocal Troullier–Martins pseudopotentials [38], Slater exchange, Perdew and Zunger correlation functionals [39], and grid spacing of 0.6 Å for a parallelepiped box of $8 \times 8 \times 60$ Å. Figure 5.16 shows the time-dependent dipoles resulting from the interaction of neutral monatomic carbon and the C_{60} molecule that is polarized in the x-axis direction parallel to the polarization direction of the electromagnetic wave with photon energy of 1.5 eV and maximum intensity of 4.8×10^{14} W cm^{-2} for 30 fs. The ions were treated as static, and the fragmentation of the C_{60} molecule was not investigated. The geometry of the C_{60} fullerene was obtained with the PCGAMESS package [40, 41]. PCGAMESS is an *ab initio* quantum chemistry software package, which was used to obtain the coordinates of the carbon atoms in the C_{60} molecule. One can see from Fig. 5.16 that for a C_{60} molecule the maximum dipole is four times larger than in monatomic carbon, which corresponds to a 16-fold enhancement of the nonlinear optical response of the generated harmonic field.

The investigation of the influence of the fundamental properties of electrons on resonant HHG can be performed by means of a multiconfigurational

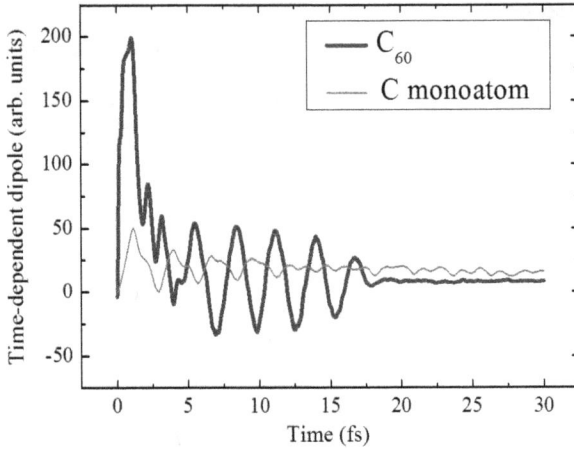

Fig. 5.16 Calculated time-dependent dipoles of C_{60} and carbon monoatom. Reprinted from [12] with permission from American Physical Society.

time-dependent Hartree–Fock approach [30, 42], which has the accuracy of direct numerical solution of the Schrödinger equation and is almost as simple as the ordinary time-dependent Hartree–Fock approach.

The MCTDHF approach treats the wavefunction of a multi-electronic system as

$$\Psi(Q_1, \ldots, Q_f, t) = \sum_{j_1=1}^{n_1} \cdots \sum_{j_f=1}^{n_f} A_{j_1 \cdots j_f}(t) \prod_{\kappa=1}^{f} \varphi_{j_\kappa}^{(\kappa)}(Q_\kappa, t) \qquad (5.10)$$

where $Q_1, \ldots, Q_f$ are the coordinates of the electrons, $A_{j_1 \cdots j_f}$ are the antisymmetrized A-vector for all n_κ time-dependent expansion functions $\phi_{j_\kappa}^{(\kappa)}$ for every degree of freedom κ. Setting $n_\kappa = n_1$ describes the direct solution of the time-dependent Schrödinger equation and $n_\kappa = 1$ simplifies the wavefunction to the ordinary time-dependent Hartree–Fock approximation.

The equations of motion in the MCTDHF approach are derived from a modified variational principle:

$$\left\langle \delta \Psi_{\mathrm{MCHF}}(t) \left| i \frac{d}{dt} - H(t) \right| \Psi_{\mathrm{MCHF}}(t) \right\rangle = 0 \forall t \qquad (5.11)$$

The MCTDHF method is applied to simulate a three-dimensional fullerene-like system represented by the so-called jelly-like sphere approximation. A jelly-like sphere is used as a potential surface for the representation of fullerenes. Then, two electrons are considered moving in this potential, while the remaining electrons are considered frozen. The system under investigation is represented by a spherically symmetric potential of the form ($R_0 = 8.1$, $R_i = 5.3$, and $v_0 = 0.78$):

$$V(r) = \begin{cases} -3 \left(\frac{250}{R_0^3 - R_i^3} \right) \left(\frac{R_0^2 - R_i^2}{2} \right), & r \leq R_i \\ - \left(\frac{250}{R_0^3 - R_i^3} \right) \left(\frac{3R_0^2}{2} - \left[\frac{r^2}{2} + \frac{R_i^3}{r} \right] \right) - v_0, & R_i < r < R_0 \\ -250/r, & r \geq R_0 \end{cases}$$

$$(5.12)$$

The Coulomb repulsion between electrons is

$$V_{ee} = 1/\sqrt{r_1^2 - r_2^2 + 2} \qquad (5.13)$$

Below, the results of a study of the interaction of the fullerene-like system with the Gaussian femtosecond electric pulse

$$E(t) = \exp \left(\frac{(t - t_0)^2}{\tau^2} \right) E_0 \sin (\omega t) \qquad (5.14)$$

where $\omega = 0.046$ atomic units (a.u., $\lambda = 991$ nm), or $\omega = 0.057$ a.u. ($\lambda = 800$ nm), are presented.

The first frequency was chosen as a source of the even (16th) harmonic, which coincides with the central region of the SPR of C_{60} ($\lambda = 60$ nm), and the second frequency coincides with the frequency of the most frequently used laser (Ti:sapphire) in such experiments.

From the solution of the MCTDHF equations for $\Psi(Q_1, \ldots, Q_f, t)$, the time-dependent dipole $d(t) = \sum_f \int \Psi(Q_1, \ldots, Q_f, t) Q_f \Psi^*(Q_1, \ldots, Q_f, t)$ was obtained for the estimation of the power spectrum of HHG. The discrete Fourier transform, which is usually used in such simulations to derive the power spectrum of HHG, suffers from inaccuracies introduced by the finiteness of the pulse because discrete Fourier transform assumes the reference signal to be infinite. Therefore, one can state that the discrete Fourier transform is in general not applicable for analysis of short signals, about three periods of a carrier wave, which one has to deal with during HHG simulations. Introduction of special window functions may improve the case, although they may introduce additional errors, which could lead to completely incorrect results. Accordingly, the HHG spectrum was analyzed by means of a piecewise least-squares-based approximation of the reference time-dependent dipole signal described as follows. During every relatively small time interval the dipole was considered as a superposition of flat waves $A_n \sin(n\omega_0 t)$ (n integer). The indices A_n are determined on the basis of a least-squares approximation and their sum over the whole signal gives the HHG spectrum. The advantage of this method is not only its full mathematic correctness and accuracy, but also the fact that the harmonic spectrum becomes easily viewable by definition.

The absorption spectrum of this system was obtained by a procedure similar to the delta-kick method and had absorption maxima near 0.741 a.u., which is the 13th harmonic of $\omega = 0.057$ a.u. and the 16th harmonic of $\omega = 0.046$ a.u. radiation, although the absorption band is rather wide (see inset in Fig. 5.17) [43]. One should mention that a simple jelly-like system

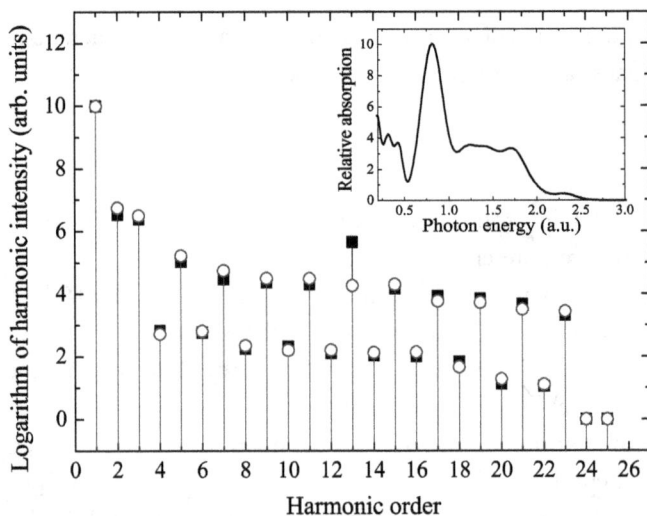

Fig. 5.17 The influence of resonance on the HHG spectrum generated in a fullerene-like medium in the cases of radiation of carrier wave frequencies 0.046 a.u. (open circles) and 0.057 a.u. (filled squares). Inset: Absorption spectrum of the fullerene-like system obtained by the delta-kick method. Reprinted from [43] with permission from American Physical Society.

was considered, not the C_{60} molecule itself, so the spectrum may deviate from the experimental one.

Figure 5.17 presents the results of HHG simulation within an exact MCT-DHF approximation (six expansion functions) for carrier wave frequencies 0.046 a.u. and 0.057 a.u. The 13th harmonic was approximately ten times enhanced relative to the plateau harmonics. Note that experimentally observed enhancement of this harmonic was approximately the same and depended on the excitation of fullerene-containing targets. One can see that harmonics neighboring the 13th are not so enhanced, although they are still close to the broad absorption band of C_{60} (50–70 nm). At the same time, they are not suppressed, so it is highly possible that a competition between enhancement and absorption takes place. The pulse in general is not monochromatic, so its spectral properties can also have an influence on the simulations.

Even harmonics were observed in the calculated spectra that were two orders of magnitude smaller than the neighboring odd harmonics. This artifact can be attributed to symmetry breaking introduced by the numerical grid, which is perhaps still too sparse and introduces some kind of rectangular integration box. Further reduction of the grid spacing may remove such unphysical results.

The resonant 13th harmonic of the radiation with carrier frequency $\omega = 0.057$ a.u. was enhanced. On the other hand, the 16th harmonic of the radiation with carrier frequency $\omega = 0.046$ a.u. was suppressed due to symmetry effects, which are still strong in the fullerene system. One can note that in both cases the maximum observed harmonic order was 23, which is close to reported experimental results at moderate excitation of fullerene-containing targets. All these results indicate that the MCTDHF approximation indeed allows us to describe both resonant HHG and harmonic cutoff in the fullerene-like medium.

Two-electron interaction is a Coulomb repulsion (Eq. 5.13) between two electrons. Neglecting the two-electron interaction destroyed the HHG process completely. At the same time, the representation of quasi-electrons as distinguishable particles had almost no influence on resonant HHG observability. The most important application of this phenomenon is the necessity for exact description of the two-electron interaction, while exchange processes have almost no effect on resonant HHG simulations.

5.7. Calculations of HHG in Endohedral Fullerenes

Resonant HHG has previously been reported to increase the efficiency of ordinary HHG up to 200 times for a single (13th) harmonic in indium plasma [19], and up to 20 times for the 11th to 15th harmonics in C_{60} plasma [3]. Resonant HHG can also lead to better control of the properties of harmonic radiation by adjusting the laser wavelength and corresponding resonant transitions. For example, in antimony plasma, a ten-fold enhancement of resonant HHG has been reported [44] using the appropriate adjustment of experimental parameters. On the other hand, HHG in fullerenes [3] has been shown to allow the simultaneous enhancement of several neighboring harmonic orders. Thus, it seems interesting to combine the superior resonant HHG properties observed in atomic plasmas with multiple harmonic enhancement in fullerene plasmas inside a single system. Here the investigations of HHG in endohedral C_{60} fullerenes with implanted indium and antimony are presented and compared with those from ordinary C_{60} fullerenes. Endohedral fullerenes [45] are characterized by the fact that electrons will transfer from the metal atom to the fullerene cage and that the metal atom takes a position off-center in the cage. In contrast, nonmetals, for example nitrogen [46], inserted in fullerenes have almost no charge transfer in the center and represent an atomic trap that is stable at room temperature and for an arbitrarily long time.

Atomic or ion traps are of great interest since particles are present free from significant interaction with their environment, allowing unique quantum mechanical phenomena to be explored, especially those which arise from quantum confinement. It is thus interesting to investigate the behavior of fullerenes doped with semiconductors, because properties of semiconductors change greatly with confinement. The endohedral C_{60} fullerene with implanted indium has not yet been obtained, but various stable endohedral fullerenes have already been obtained experimentally, not only La@C_{60}. In [47], the formation of heavier atom (Sb, Te)-incorporated fullerenes has been investigated by using radionuclides produced by nuclear reactions. From a trace of the radioactivities of ^{120}Sb (^{122}Sb) or ^{121}Te after high-performance liquid chromatography, it was found that the formation of endohedral fullerenes or heterofullerenes in atoms of antimony or tellurium is possible by a recoil process following the nuclear reactions. The results of a molecular dynamics simulation in [47] can also easily describe the ion implantation process, which can substitute the nuclear recoil previously used to implant neutral atoms. Similar experimental and theoretical results were presented in [48] for a broader range of targets.

Endohedral fullerenes can also be produced by arc-discharge vaporization of composite rods made of graphite and metal compounds [49, 50], although the output is relatively low. There are also techniques for insertion of foreign atoms into the C_{60} shell [51–53]. But taking into account the conditions of typical HHG experiments, an optimal and inexpensive technique to produce them could be ion implantation during laser ablation of C_{60}-containing superglue, because expensive purification as well as long-term stability of endohedral fullerenes would not be needed. Although the optimal conditions for the experiments should be investigated, production of In@C_{60} and Sb@C_{60} in quantities sufficient for HHG experiments is possible in principle. The macroscopic stability of C_{60} with some implanted ions is unlikely. However, this is actually not required, because the duration of any HHG experiment is shorter than the decay time of endoheral fullerenes obtained by means of ion implantation into vaporized fullerene molecules, so there is a high possibility of experimental verification of this effect by adding an ion beam into the standard HHG setup.

Spectroscopic data for endohedral fullerenes in the XUV range are also not yet available. The investigation described below is devoted to the spectroscopic properties of C_{60} doped with indium and antimony and their effect on HHG in the case of 800 nm fundamental laser radiation [54]. The theoretical investigation of absorption and HHG spectra of C_{60}, In@C_{60}, and Sb@C_{60} and a study of their possible applications are performed.

Because HHG requires a field strength of the pump laser radiation close to the intra-atomic one, the laser field can no longer be treated as a small perturbation, and time-independent methods, such as perturbation theory, are not applicable. To model the HHG one has to solve numerically the time-dependent Schrödinger equation in the TDDFT approximation (see previous section) with the aid of real-space real-time code. Detailed description of TDDFT formalism can be found in [55].

The adiabatic local density approximation most widely utilized in the literature is not very useful when describing the HHG process, which is strongly nonadiabatic. For all the calculations here we used the Krieger–Li–Iafrate method [56] applicable for time-dependent problems. Here, the exchange-correlation potential V_{xc}^{KLI} is expressed as a set of explicit functionals of the orbitals $\varphi_{j\sigma}$, where σ stands for the spin of the electrons. In the exchange-only case V_x^{KLI}, if correlation effects are neglected, the potential reads

$$V_x^{KLI}(\vec{r}, t) = w_{x\sigma}(\vec{r}, t) + \frac{1}{\rho_\sigma(\vec{r}, t)} \sum_j^{n_\sigma} \rho_{j\sigma}(\vec{r}, t) \times \int d^3\vec{r}' \rho(\vec{r}', t) V_x^{KLI}(\vec{r}', t)$$

(5.15)

where $\rho_{j\sigma}(\vec{r}, t) = |\varphi_{j\sigma}(\vec{r}, t)|^2$ and

$$w_{x\sigma}(\vec{r}, t) = -\frac{1}{\rho_\sigma(\vec{r}, t)} \sum_{j,k}^{n_\sigma} \phi_{j\sigma}(\vec{r}, t) \phi_{k\sigma}^*(\vec{r}, t) \int d^3\vec{r}'$$

$$\times \frac{\phi_{k\sigma}(\vec{r}, t) \phi_{j\sigma}^*(\vec{r}, t)}{|\vec{r} - \vec{r}'|} - \rho_{j\sigma}(\vec{r}, t) \int d^3\vec{r}'' \int d^3\vec{r}'$$

$$\times \frac{\phi_{k\sigma}(\vec{r}, t) \phi_{j\sigma}^*(\vec{r}, t) \phi_{j\sigma}(\vec{r}, t) \phi_{k\sigma}^*(\vec{r}, t)}{|\vec{r}''\vec{r}'|}$$

(5.16)

Before the time-dependent runs, self-consistent ground states were obtained for all the investigated systems by minimization of the overall energy of electronic orbitals $\varphi_i(\vec{r}, t)|_{t=0}$ to some convergence criteria. C_{60} geometry was obtained from freely available results of geometry optimization by means of molecular dynamics. By constructing a full wavefunction for endohedral fullerenes, there is no atom–fullerene wavefunction overlap at all, because these studies seek to model endohedral fullerenes as being produced by means of ion implantation during laser ablation. So the self-consistent wavefunctions of fullerene shells and endohedral doping are simply added to each other and then renormalized. The numerical results show that under such circumstances the influence of screening is insignificant; that is why plasmon resonance tunes towards stronger atomic resonances.

Pseudopotentials [57] were used for carbon, indium, and antimony in endohedral fullerenes to reduce the total number of electrons. In@C_{60} and Sb@C_{60} are open-shell systems, so additional electronic orbitals were added to account for the lowest excited states (in the computations, three excited states were obtained from the properties of the potential without additional assumptions) and achieve convergence. They were also added for ordinary

C_{60} for comparison. To decrease numerical effort without a loss of sense, the simulation box was a parallelepiped, 80 a.u. along the propagation axis and 20 a.u. on the other two axes. Grid spacing was 0.5 a.u. One should note that the nonspherical integration box violates the symmetry of the system, so even harmonics may appear in the calculations, while they should not exist in the experiment under such conditions.

Harmonic generation in three media (C_{60}, In@C_{60}, and Sb@C_{60}) was analyzed by applying an electric field corresponding to the commonly used laser source (Ti:sapphire laser) with a central wavelength of 800 nm and a pulse duration of 48 fs. It is well known for HHG in real fullerenes that electrons from the lowest energy levels are unlikely to ionize, so it is quite justified to consider them "frozen." Only the evolution of 8 of the 125 electronic orbitals was computed, although the action of Hartree and exchange potentials of these frozen orbitals on the investigated ones was fully considered. In the time-dependent runs, density reaching the boundary of the integration box was removed by a mask technique similar to that in [58]. To obtain the photoabsorption spectrum of the studied systems the delta-kick method [59] with the same simulation region was used. This was also a good chance to check the correspondence of the studied model system to the real one. This means a strong short delta-shaped pulse was applied at the beginning of the simulation, after which no external field was applied to the system so that density propagated freely to the ground state during the first 48 fs with a timestep of 0.04 a.u. (0.001 fs). Then, propagation under the influence of a strong pulse with an amplitude of 0.5 a.u. and a frequency of 0.057 a.u. ($\lambda = 800$ nm) broadened by a cosinusoidal envelope $\cos{(\pi(t - 2\tau_0 - t_0)/\tau_0)}$, where $\tau_0 = 0.5$ and $t_0 = 1200$ (peak of the contour), was investigated under the same conditions for the first 48 fs.

Introduction of pseudopotentials for a multi-electron system with corresponding exchange functionals eliminates the need for forced specification of artificial autoionizing resonances, and the time propagation methods of spectrum investigation do not require the addition of unoccupied states. Hence the spectrum is derived *ab initio* in a natural way for a given model system.

The square of the absorption cross-sections of all species as a function of the photon energy in units of 0.057 a.u., which corresponds to the harmonic orders of 800 nm laser radiation, is presented in Figs. 5.18a–5.18c. There are signs of separate resonant peaks whose experimental observations were

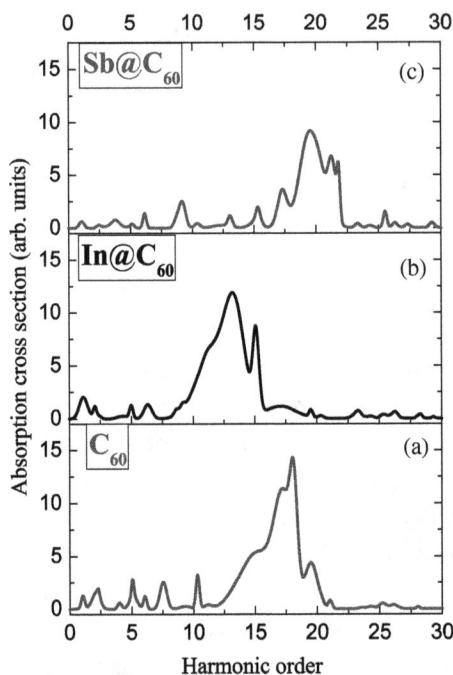

Fig. 5.18 Absorption spectra of (a) original C_{60} fullerene, (b) endohedral In@C_{60} fullerene, and (c) endohedral Sb@C_{60} fullerene. Energy is given in harmonics of fundamental radiation with frequency 0.057 a.u. Reprinted from [54] with permission from American Physical Society.

previously attributed to strong transitions in indium and antimony [60, 61]. This may be because the C_{60} shell does not prevent the corresponding dipole excitations of implanted atoms. More important is the fact that broad plasmon peaks are shifted from the energy of 17th harmonic of 800 nm radiation in ordinary C_{60} towards the 13th and 19th to 21st harmonics in In@C_{60} and Sb@C_{60} respectively.

Due to the nature of the TDDFT formalism it is impossible to distinguish which electrons are responsible for the observed spectral shifts. However, it is now evident that even in the absence of charge transfer, the valence electrons of the implanted atoms are not screened from the pump radiation, because even a slight excitation of other electrons is equal to perturbations similar to a delta-kick, and the system of delocalized electrons moves as a whole. So any noticeable resonance in atoms is actually amplified by the collective oscillation modes of the C_{60} shell, but the resulting oscillation is in turn a seed for collective oscillations because its energy is mostly absorbed by collective oscillation modes, which are closer to the corresponding resonance in implanted media. That is why the collective oscillations of delocalized electrons can tune to such resonances. As far as the equilibrium conditions are determined by the stability of collective oscillation modes, the initial strength of the resonance plays almost no role in the strength of the resulting spectral properties.

The results of simulations of the HHG process in the three studied fullerenes are presented in Figs. 5.19a–5.19c. It should be noted that the results are plotted on a nonlogarithmic scale where the intensity of the not-shown first harmonic is 10,000, so the resonantly enhanced harmonics had intensity up to 10^{-4} of the main pulse. Real conversion efficiency is usually lower by an order of magnitude due to self-defocusing on free electrons during propagation. One can easily see a good correspondence between spectral

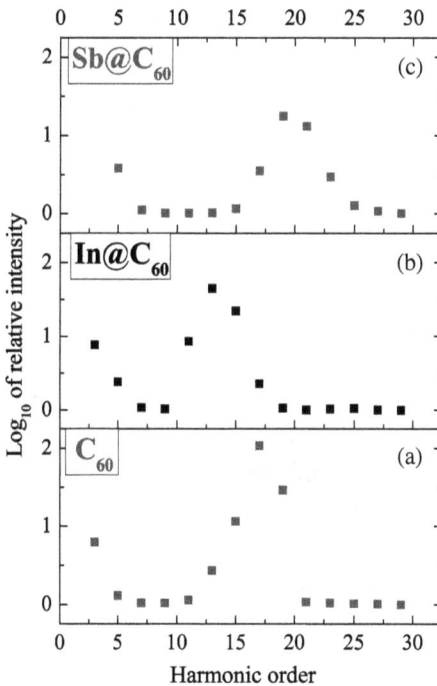

Fig. 5.19   HHG spectra from (a) original C_{60} fullerene, (b) endohedral In@C_{60} fullerene, and (c) endohedral Sb@C_{60} fullerene. Intensities are normalized so that the intensity of the first harmonic (i.e., fundamental intensity) is 10,000 (not shown). Reprinted from [54] with permission from American Physical Society.

properties (Fig. 5.18) and positions of enhanced groups of harmonics presented on Fig. 5.19. A very important observation comes from the fact that the shifted peaks of HHG enhancement in endohedral In@C_{60} and Sb@C_{60} are very close to the corresponding shifted absorption peaks in them. It may be noted that in the case of pure fullerenes maximum intensity of harmonics was observed in the vicinity of the broad SPR of C_{60} with the central wavelength near 50 nm (i.e., close to the 15th harmonic of an 800 nm pump).

The enhancement of groups of harmonics was analyzed in previous sections during HHG in fullerene-containing plasma plumes. The enhanced harmonics were observed in the range of the 13th harmonic, which was attributed to the influence of the SPR of neutral C_{60} ($\lambda_{SPR} \cong 60$ nm with 10 nm full width at half maximum).

Single harmonic enhancement using indium and antimony plumes has also been experimentally observed. Using indium laser-ablation, enhancement of the 13th harmonic at the wavelength of 61.26 nm has been reported [1]. The $4d^{10} 5s^{21} S_0 \rightarrow 4d^9 5s^2 5p (^2D)\ ^1P_1$ transition of In II at the wavelength of 62.2 nm, which has an oscillator strength (gf value) of 1.11, can be driven into resonance with the 13th harmonic by the AC Stark shift. The gf value of this transition is ten times higher than for the other transitions of In II in this spectral range. In the case of antimony plasma, enhancement of the 21st harmonic of the 791 nm radiation was reported [44], although the 21st harmonic ($\lambda = 37.67$ nm) was slightly away from the $4d^{10} 5s^2 2p^3 P_2 \rightarrow 4d^9 5s^2 5p^3 (^2D)\ ^3D_3$ transition ($\lambda = 37.82$ nm) possessing the highest gf value.

Thus, the experimentally observed enhanced harmonics in the cases of C_{60} (13th harmonic), indium (13th harmonic), and antimony (21st harmonic) were close to the enhanced harmonics calculated in the present study (17th harmonic in the case of pure fullerenes, 13th harmonic in the case of In@C_{60}, and 19th harmonic in the case of Sb@C_{60}). The difference between experiment and theory in the case of pure fullerenes could be related to the conditions of the experiment, where the plasma contained mostly neutral fullerenes, for which SPR lies in the range of 60 nm (i.e., the 13th harmonic of laser radiation).

There are also low-energy absorption peaks in the absorption spectra of all the systems. So it is interesting to find out why resonant harmonic generation was observed only for those in the high-energy domain. A possible explanation of this phenomenon can be given by approaches that attribute resonant HHG to resonant properties of the media at the moment of recombination [43, 62].

Namely, resonant recombination can proceed through every resonance, but when the electron moves back to the nucleus, the recombination into the first available resonant channel (with the highest energy) is extremely probable due to either its sufficiently large matrix dipole element [43] or inelastic collision cross-section [62]. So, a strong resonant transition in the XUV range is a better choice for resonant HHG than the same transition in the lower energy range. Phase matching conditions are in general worse for higher harmonics because of self-defocusing on free electrons, so in the experiments, the lower intensity of higher harmonics is a result of the propagation effect, which is not included in our computations.

No theoretical or experimental evidence of HHG shifts in endohedral fullerenes has been reported yet. If the presented results are experimentally verified, they could open up a new way to control resonant HHG. This is only a matter of time, because there has already been experimental success in resonant HHG in indium and antimony plasmas as well as enhancement of groups of harmonics near SPR of C_{60}. There is also no counter-evidence of the possibility of formation of stable $In@C_{60}$. In addition, $Sb@C_{60}$ is stable and available in quantities sufficient for experiments. The approach proposed here can be used to increase the strength of groups of harmonics near SPR of C_{60}. Further advances in this area of research include the use of ionized C_{60} shells, which could promote the peak of HHG enhancement deeper into the far UV spectral range, from 20 eV for neutral C_{60} to 25 eV for singly ionized C_{60}.

Below we briefly address the possibility of generating attosecond pulses in the fullerenes under consideration. It has been discussed in [62] that resonant enhancement of a single harmonic is insufficient for production of attosecond pulse trains. The situation is even worse in a mixture of them, because resonant conditions for certain types of atoms differ greatly because of their narrow absorption peaks. The suggestion [54] to use a mixture of endohedral fullerenes as a medium for attosecond pulse train generation is only a development of the suggestions first proposed in [62], where atomic targets with close resonances were predicted to be useful media for attosecond pulse train generation. In fact, it was shown in [62] that even a single resonant HHG is sufficient to generate a bright attosecond pulse train, but its shape is too poor due to domination of a single resonant harmonic. In C_{60}, neighboring harmonics present a better case, since their intensities in the range of SPR have relatively equal values. As far as resonant harmonics from endohedral fullerenes have the same plasmon-enhanced origin, we propose that a simple mix of ordinary C_{60} and endohedral fullerenes could produce the required intensity distribution.

Several resonantly enhanced harmonics in C_{60} and endohedral fullerenes could be a possible solution to this problem, because the corresponding phases of resonant harmonics are linked to each other [12, 63]. For the best attosecond pulse train production these neighboring harmonics should be of comparable intensity. In [63], an atom with several transitions at suitable frequencies and similar oscillator strengths, or a mixture of atoms, was suggested for this purpose. Unfortunately, it turns out to be impossible to find corresponding atoms with such transitions for which resonant conditions could be fulfilled by a single probe pulse.

5.8. Discussion

We now discuss the origin of the enhancement of HHG in C_{60}. Specifically, we focus on two important observations, namely (i) extension of harmonic cutoff and (ii) enhancement of harmonics in the vicinity of plasmon resonance. In the three-step model for HHG, the cutoff harmonic is given by $I_p + 3.17 \times U_p$ (where $I_p = 7.6\,\text{eV}$ is the ionization potential of C_{60} and U_p is the ponderomotive energy). An intensity of $\sim 10^{14}\,\text{W cm}^{-2}$ was used in the measurements, which is above the saturation intensity of the first two charge states of C_{60} [6]. The experimentally measured saturation intensity of C_{60}^+ is $5 \times 10^{13}\,\text{W cm}^{-2}$, in close agreement with the theoretical value of $4 \times 10^{13}\,\text{W cm}^{-2}$ [8]. Accordingly, if the HHG is from neutral C_{60} the cutoff should be at the 11th harmonic. In contrast, harmonics above 19th order at small intensity of the driving pulse were observed.

Higher cutoff could be due to (i) the contribution of C_{60} ions to the HHG process, since laser ablation at the pulse intensities used in the experiment is known to lead to soft ionization identical to matrix-assisted laser desorption/ionization; (ii) multiphoton excitation of surface plasmons (20 eV) by the incident laser field (1.55 eV). If ionization starts from a plasmon state and the electron returns to the ground state upon recombination, the plasmon energy is converted into photon energy extending the cutoff. (iii) Recombination into orbitals, other than the highest occupied molecular orbital of C_{60}, with higher ionization potentials can result in extension of cutoff.

High harmonics in C_{60} or in any complex multi-electron system will have two contributions — the usual harmonic generation process and the physical mechanisms that lead to enhancement of harmonics (9th to 15th in C_{60}) around the frequencies at which the system displays collective electron oscillations (20 eV in C_{60} with a full width at half maximum of ~ 10 eV). Plasmon excitation under two different scenarios can lead to enhancement of high harmonics. (i) The recolliding electron excites the plasmon upon recombination, which then decays by emitting photons. This leads to enhancement of the harmonics in the vicinity of plasmon resonance [9–30]. Such a mechanism would be wavelength dependent. (ii) The laser field directly excites the surface plasmons through a multiphoton process, which then decay back by emitting high-energy photons. Similar bound–bound transitions have been theoretically shown to exist in C_{60}. Such a mechanism would be independent of the recollision process and could be revealed by ellipticity measurements.

The higher harmonic yield in a highly polarizable molecule such as C_{60} compared to an atom could simply be due to enhanced recombination cross sections resulting from its larger spatial extent. Also, the delocalized electron distribution can lead to a large induced dipole, as shown above. The harmonic efficiency depends on the square of the dipole matrix. While these effects explain higher harmonic yield in C_{60} in general, they also provide insight into why only harmonics near SPR are enhanced.

References

1. Ganeev, R.A., Suzuki, M., Ozaki, T. *et al.* (2006). Strong resonance enhancement of a single harmonic generated in extreme ultraviolet range, *Opt. Lett.*, **31**, 1699–1701.
2. Shim, B., Hays, G., Zgadzaj, R. *et al.* (2007). Enhanced harmonic generation from expanding clusters, *Phys. Rev. Lett.*, **98**, 123902.
3. Ganeev, R.A., Elouga Bom, L.B., Abdul-Hadi, J. *et al.* (2009). High-order harmonic generation from fullerene using the plasma harmonic method, *Phys. Rev. Lett.*, **102**, 013903.
4. Hertel, I.V., Steger, H., de Vries, J. *et al.* (1992). Giant plasmon excitation in free C_{60} and C_{70} molecules studied by photoionization, *Phys. Rev. Lett.*, **68**, 784–787.
5. Scully, S.W.J., Emmons, E.D., Gharaibeh, M.F. *et al.* (2005). Photoexcitation of a volume plasmon in C_{60} ions, *Phys. Rev. Lett.*, **94**, 065503.
6. Bhardwaj, V.R., Rayner, D.M. and Corkum, P.B. (2003). Internal laser-induced dipole force at work in C_{60} molecule, *Phys. Rev. Lett.*, **91**, 203004.

7. Bhardwaj, V.R., Rayner, D.M. and Corkum, P.B. (2004). Recollision during the high laser intensity ionization of C_{60}, *Phys. Rev. Lett.*, **93**, 043001.

8. Jaron-Becker, A., Becker, A. and Faisal, F.H.M. (2006). Saturated ionization of fullerenes in intense laser fields, *Phys. Rev. Lett.*, **96**, 143006.

9. Hoshi, H., Manaka, T., Ishikawa, K. *et al.*(1997). Second-harmonic generation in C_{70} film, *Jpn. J. Appl. Phys.*, **36**, 6403–6404.

10. Neher, D., Stegeman, G.I., Tinker, F.A. *et al.* (1992). Nonlinear optical response of C_{60} and C_{70}, *Opt. Lett.*, **17**, 1491–1493.

11. Kafafi, Z.H., Lindle, J.R., Pong, R.G.S. *et al.* (1992). Off-resonant nonlinear optical properties of C_{60} studied by degenerate four-wave mixing, *Chem. Phys. Lett.*, **188**, 492–496.

12. Ganeev, R.A., Elouga Bom, L.B., Wong, M.C.H. *et al.* (2009). High-order harmonic generation from C_{60}-rich plasma, *Phys. Rev. A*, **80**, 043808.

13. Ganeev, R.A., Singhal, H., Naik, P.A. *et al.* (2009). Influence of C_{60} morphology on high-order harmonic generation enhancement in fullerene-containing plasma, *J. Appl. Phys.*, **106**, 103103.

14. Ganeev, R.A., Baba, M., Kuroda, H. *et al.* (2011). Low- and high-order nonlinear optical characterization of C_{60}-containing media, *Eur. Phys. J. D*, **64**, 109–114.

15. Ganeev, R.A., Baba, M., Suzuki, M. *et al.* (2005). High-order harmonic generation from silver plasma, *Phys. Lett. A*, **339**, 103–109.

16. Rubenchik, A.M., Feit, M.D., Perry, M.D. *et al.* (1998). Numerical simulation of ultrashort laser pulse energy deposition and bulk transport for material processing, *Appl. Surf. Sci.*, **129**, 193–198.

17. Wulker, C., Theobald, W., Ouw, D. *et al.* (1994). Short-pulse laser-produced plasma from C_{60} molecules, *Opt. Commun.* **112**, 21–28.

18. Shchatsinin, I., Laarmann, T., Zhavoronkov, N. *et al.* (2008). Ultrafast energy redistribution in C_{60} fullerenes: A real time study by two-color femtosecond spectroscopy, *J. Chem. Phys.*, **129**, 204308.

19. Ganeev, R.A., Singhal, H., Naik, P.A. *et al.* (2006). Harmonic generation from indium-rich plasmas, *Phys. Rev. A*, **74**, 063824.

20. Ganeev, R.A. (2007). High-order harmonic generation in laser plasma: A review of recent achievements. *J. Phys. B*, **40**, R213–R253.

21. Charalambidis, D., Tzallas, P., Benis, E.P. *et al.* (2008). Exploring intense attosecond pulses, *New J. Phys.*, **10**, 025018.

22. Ganeev, R.A., Singhal, H., Naik, P.A. *et al.* (2010). Enhanced harmonic generation in C_{60}-containing plasma plumes, *Appl. Phys. B*, **100**, 581–585.

23. Ganeev, R.A., Singhal, H., Naik, P.A. *et al.* (2009). Enhancement of high-order harmonic generation using two-color pump in plasma plumes, *Phys. Rev. A*, **80**, 033845.

24. Ganeev, R.A., Suzuki, M., Baba, M. *et al.* (2008). High-order harmonic generation in Ag nanoparticle-containing plasma, *J. Phys B: At. Mol. Opt. Phys.*, **41**, 045603.

25. Ganeev, R.A. (2008). High-order harmonic generation in nanoparticle-containing laser-produced plasmas, *Laser Phys.*, **18**, 1009–1015.

26. Ganeev, R.A., Suzuki, M., Baba, M. *et al.* (2008). Low- and high-order nonlinear optical properties of $BaTiO_3$ and $SrTiO_3$ nanoparticles, *J. Opt. Soc. Am. B*, **25**, 325–333.

27. Ganeev, R.A., Chakravarty, U., Naik, P.A. *et al.* (2007). Pulsed laser deposition of metal films and nanoparticles in vacuum using subnanosecond laser pulses, *Appl. Opt.*, 46, 1205–1210.
28. Zhu, W., Miser, D.E., Chan, W.G. *et al.* (2004). Characterization of combustion fullerene soot, C_{60}, and mixed fullerene, *Carbon*, 42, 1463–1471.
29. Ciappina, M.F., Becker, A. and Jaron-Becker, A. (2007). Multislit interference patterns in high-order harmonic generation in C_{60}, *Phys. Rev. A*, 76, 063406.
30. Cricchio, D., Corso, P.P., Fiordilino, E. *et al.* (2009). A paradigm of fullerene, *J. Phys. B: At. Mol. Opt. Phys.*, 42, 085404.
31. Zhang, G.P. (2005). Optical high harmonic generation in C_{60}, *Phys. Rev. Lett.*, 95, 047401.
32. Constant, E., Garzella, D., Breger, P. *et al.* (1999). Optimizing high harmonic generation in absorbing gases: Model and experiment, *Phys. Rev. Lett.*, 82, 1668–1671.
33. Windt, D.L., Cash, W.C., Scott, M. *et al.* (1988). Optical constants for thin-films of C, diamond, Al, Si, and CVD SiC from 24 Å to 1216 Å, *Appl. Opt.*, 27, 279–285.
34. Henke, B.L., Gullikson, E.M. and Davis, J.C. (1993). X-Ray interactions: Photoabsorption, scattering, transmission, and reflection at $E = 50$–$30{,}000\,\mathrm{eV}$, $Z = 1$–92, *At. Data Nucl. Data Tables*, 54, 181–342.
35. Mori, T., Kou, J., Haruyama, Y. *et al.* (2005). Absolute photoabsorption cross section of C_{60} in the extreme ultraviolet, *J. Electron Spectrosc. Relat. Phenom.*, 144, 243–246.
36. Runge, E. and Gross, E.K.U. (1984). Density-functional theory for time-dependent systems, *Phys. Rev. Lett.*, 52, 997–1000.
37. Marques, M.A.L., Castro, A., Bertsch, G.F. *et al.* (2003). Octopus: A first-principles tool for excited electron–ion dynamics, *Comp. Phys. Commun.*, 151, 60–78.
38. Troullier, N. and Martins, J.L. (1991). Efficient pseudopotentials for plane-wave calculations, *Phys. Rev. B*, 43, 1993–2006.
39. Perdew, J.P. and Zunger, A. (1981). Self-interaction correction to density-functional approximations for many-electron systems, *Phys. Rev. B*, 23, 5048–5079.
40. Schmidt, M.W., Baldridge, K.K., Boatz, J.A. *et al.* (1993). General atomic and molecular electronic structure system, *J. Comput. Chem.*, 14, 1347–1363.
41. Nemukhin, A.V., Grigorenko, V.L. and Granovsky, A.A. (2004). Molecular modeling by using the PC GAMESS program from diatomic molecules to enzymes, *Moscow Univ. Chem. Bull.*, 45, 75–101.
42. Zanghellini, J., Jungreuthmayer, C. and Brabec, T. (2006). Plasmon signatures in high harmonic generation, *J. Phys. B: At. Mol. Opt. Phys.*, 39, 709–718.
43. Redkin, P.V. and Ganeev, R.A. (2010). Simulation of resonant high-order harmonic generation in three-dimensional fullerenelike system by means of multiconfigurational time-dependent Hartree–Fock approach, *Phys. Rev. A*, 81, 063825.
44. Suzuki, M., Baba, M., Kuroda, H. *et al.* (2007). Intense exact resonance enhancement of single-high-harmonic from an antimony ion by using Ti:sapphire laser at 37 nm, *Opt. Express*, 15, 1161–1166.
45. H. Shinohara, (2000). Endohedral metallofullerenes, *Rep. Prog. Phys.*, 63, 843–892.
46. Saunders, M., Jiménez-Vázquez, H.A., Cross, R.J. *et al.* (1993). Stable compounds of helium and neon: He@C_{60} and Ne@C_{60}, *Science*, 259, 1428–1430.
47. Ohtsuki, T., Ohno, K., Shiga, K. *et al.* (2001). Formation of Sb- and Te-doped fullerenes by using nuclear recoil and molecular-dynamics simulations, *Phys. Rev. B*, 64, 125402.

48. Ohtsuki, T. and Ohno, K. (2004). Radiochemical approaches for formation of endohedral fullerenes and MD simulation, *Sci. Technol. Adv. Mater.*, **5**, 621–624.
49. Y. Chai, T. Guo, C. Jin, *et al.* (1991). Fullerenes with metals inside, *J. Phys. Chem.*, **95**, 7564–7568.
50. Takata, M., Umeda, B. Nishibori, E. *et al.* (1995). Confirmation by X-ray diffraction of the endohedral nature of the metallofullerene Y@C_{82}, *Nature*, **377**, 46–49.
51. Saunders, M., Cross, R.J. Jimenez-Vazquez, H.A. *et al.* (1996). Noble gas atoms inside fullerenes, *Science*, **271**, 1693–1697.
52. Braun, T. and Rausch, H. (1998). Radioactive endohedral metallofullerenes formed by prompt gamma-generated nuclear recoil implosion, *Chem. Phys. Lett.*, **288**, 179–182.
53. Gadd, G.E., Schmidt, P., Bowles, C. *et al.* (1998). Evidence for rare gas endohedral fullerene formation from γ recoil from HPLC studies, *J. Am. Chem. Soc.*, **120**, 10322–10325.
54. Redkin, P.V., Danailov, M. and Ganeev, P.A. (2011). Endohedral fullerenes: A way to control resonant high-order harmonic generation, *Phys. Rev. A*, **84**, 013407.
55. Marques, M.A.L., Ullrich, C.A., Nogueira, F. *et al.* (2003). *Time-Dependent Density Functional Theory*, Springer-Verlag, Berlin–Heidelberg.
56. Krieger, J.B., Li, Y. and Iafrate, G.J. (1992). Construction and application of an accurate local spin-polarized Kohn–Sham potential with integer discontinuity: Exchange-only theory, *Phys. Rev. A*, **45**, 101–126.
57. Hartwigsen, C., Goedecker, S. and Hutter, J. (1998). Relativistic separable dual-space Gaussian pseudopotentials from H to Rn, *Phys. Rev. B*, **58**, 3641–3662.
58. Tafipolsky, M. and Schmid, R. (2006). A general and efficient pseudopotential Fourier filtering scheme for real space methods using mask functions, *J. Chem. Phys.*, **124**, 174102.
59. Bertsch, G.F., Iwata, J.I., Rubio, A. *et al.* (2000). Real-space, real-time method for the dielectric function, *Phys. Rev. B*, **62**, 7998–8002.
60. Duffy, G. and Dunne, P. (2001). The photoabsorption spectrum of an indium laser produced plasma, *J. Phys. B: At. Mol. Opt. Phys.*, **34**, L173–L178.
61. D'Arcy, R., Costello, J.T., McGuinnes, C. *et al.* (1999). Discrete structure in the 4d photoabsorption spectrum of antimony and its ions, *J. Phys. B: At. Mol. Opt. Phys.*, **32**, 4859–4876.
62. Strelkov, V. (2010). Role of autoionizing state in resonant high-order harmonic generation and attosecond pulse production, *Phys. Rev. Lett.*, **104**, 123901.
63. Antoine, P., L'Huillier, A. and Lewenstein, M. (1996). Attosecond pulse trains using high-order harmonics, *Phys. Rev. Lett.*, **77**, 1234–1237.

6

Enhancement of Harmonic Yield from Ablation Plumes

In this chapter, we consider various methods for increasing the harmonic yield during HHG from plasma plumes. Among them are the two-color pump method for enhancement of harmonic output from plasma in the whole plateau region, application of time-resolved spectroscopy of laser plasma for enhancement of harmonic efficiency and generation of a second plateau in the harmonic distribution, as well as the use of carbon aerogel plumes as efficient media for HHG in the 40–90 nm range. We also present comparative studies of HHG in laser plasma and gases, demonstrating the advanced properties of this method of frequency conversion, which shows that under some conditions the conversion efficiency from plasma can exceed that from gases.

6.1. Two-Color Pump for Enhancement of Harmonic Output from Plasma over the Whole Plateau Region

As already mentioned, the joint application of the fundamental (ω) and the second-harmonic (2ω) pumps led to harmonic enhancement in gases [1–7]. The spectral and temporal structure of high harmonic emission from argon exposed to an infrared laser field and its second-harmonic have been investigated in [4]. For a wide range of harmonic-generation conditions, trains of attosecond pulses with only one pulse per infrared cycle were generated. The synchronization necessary for producing such trains ensures that they have a stable pulse-to-pulse carrier envelope phase, unlike trains generated from single-color fields, which have two pulses per cycle and a π phase shift between consecutive pulses. A method of producing single attosecond pulses by HHG with multicycle driver laser pulses is reported in [5]. The driving pulse is tailored so that attosecond pulses are produced only every full cycle of the oscillating laser field, rather than every half-cycle. Because the two-color laser field contains more control parameters than a single laser field, an optimization of the experimental parameters is required for strong and efficient HHG in the two-color laser field. Very efficient HHG in a two-color laser field using a long helium gas jet has been reported [3, 6]. With the optimization of laser parameters and target conditions, strong harmonics were produced at $2(2n + 1)$th orders in an orthogonally polarized two-color field.

Some applications of a two-color pump for plasma HHG were described in previous chapters. Below, we address this technique in detail. In [8], the first observation was reported of a large increase in the harmonic conversion efficiency from plasma plumes irradiated by an intense two-color femtosecond laser beam, in which the fundamental field and its weak SH (energy ratio of 50:1) were linearly polarized orthogonal to each other. It became obvious that this technique of harmonic enhancement is a very useful tool for harmonic studies in plasma plumes. The next set of these studies of two-color pump-induced HHG in plasmas [9] is analyzed below. It was shown in [9] that the increase of second-harmonic intensity allows the generation of only even harmonics ($2n\omega$ and not just $2(2n + 1)\omega$), when odd high-order harmonics are almost suppressed. These observations were obtained in the case of a 12:1 energy ratio between the fundamental and second-harmonic waves. Overall, it was shown that the use of a two-color pump in laser-produced plasma plumes is a versatile technique for efficient odd and even harmonic generation, and can serve as a potential source of attosecond pulses.

Below we show one example of observation of harmonic enhancement in the two-color pump scheme, which was demonstrated for gold nanoparticle-containing plasma. The experimental scheme was analogous to

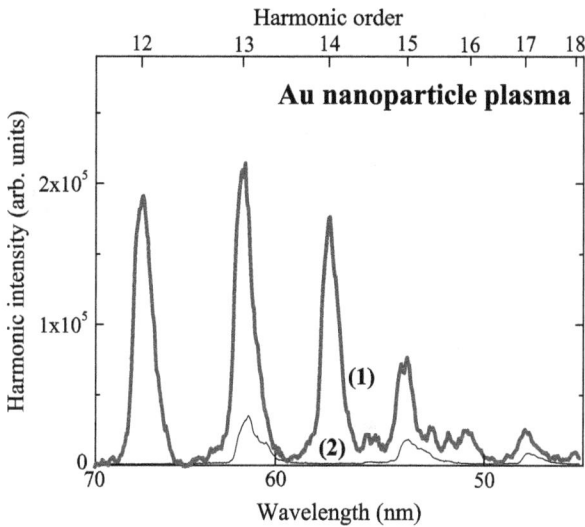

Fig. 6.1 Harmonic spectra obtained from gold nanoparticle-containing plasma for (1) two-color pump and (2) single-color pump. Reprinted from [9] with permission from American Physical Society.

that shown in Section 3.4. Figure 6.1 presents the comparative spectra of the harmonics generated using the single-color (curve 2) and two-color (curve 1) fields in a plasma containing gold nanoparticles. One can see considerable enhancement of the harmonic yield in the latter case. The gold nanoparticles (15–30 nm diameter) showed mostly spherical shape, while in some cases a triangular form of these clusters was observed. The absorption peak of these nanoparticles was around 530 nm, corresponding to the commonly reported SPR of gold spherical clusters. The nanoparticles were glued onto a glass substrate and ablated by picosecond pulses prior to the ω and 2ω waves arriving into the nanoparticle-containing plasma. To achieve efficient harmonic generation from the nanoparticles, one should not over-excite the nanoparticle-containing targets. The excitation should be just sufficient to produce a plasma plume where these species are present in their indigenous neutral (or ionized) condition. Note that these harmonics were considerably stronger than those generated in the single particle-containing plasma created on the surface of bulk gold. The nanoparticle-containing plasma plumes demonstrated a considerable enhancement of the low-order harmonics in the plateau region and a decrease of harmonic cutoff compared to the monoparticle-containing plasmas produced on the surface of bulk targets of the same origin.

One of the objectives of this study was to define whether one could increase the harmonic cutoff and conversion efficiency of HHG from gold nanoparticles using a two-color pump compared to the single-color pump. The cutoff of the harmonic spectrum did not increase; however, enhancement in the intensity of the odd harmonics was observed with two-color laser pulses. Though the conversion efficiency of the second-harmonic pump was quite low (8%), the intensity of the even harmonics was comparable to that of the odd harmonics. The intensity of the even harmonics decreased more rapidly than that of the odd harmonics, with increasing harmonic orders. Utilization of a two-color pump results in an increase in overall conversion efficiency of the HHG process, since the conversion of even harmonic orders is now added, with the odd harmonic orders showing stronger yield. In this case, no harmonic generation was observed by filtering the fundamental radiation and using only a second-harmonic pump, thus indicating that the intensity of the second-harmonic pulse itself was not sufficient for HHG. This behavior shows that even a small amount of second-harmonic photons was sufficient to break the inversion symmetry and cause both even and odd harmonic generation.

The intensity of the odd harmonics is expected to be considerably stronger than that of the even harmonics, taking into account the relative intensities of the fundamental and second-harmonic waves. However, the relative phase between these pumps is observed to change the ratio between the odd and even harmonics considerably. At appropriate conditions, this ratio can be considerably diminished, especially in the shorter wavelength range.

In Fig. 6.2, the spectra of the harmonics generated in a silver plasma plume using the single-color and two-color pump schemes are shown. One can see that in the case of the single-color pump, a conventional pattern of odd harmonics dominates the spectral pattern (thin curve). The lobes on both sides of the harmonics represent the second-order diffraction pattern obtained from the grating. Using the two-color scheme, different ratios of the odd and even harmonics appear, depending on the conditions of the experiments, in particular on the relative intensity of the second-harmonic wave and the

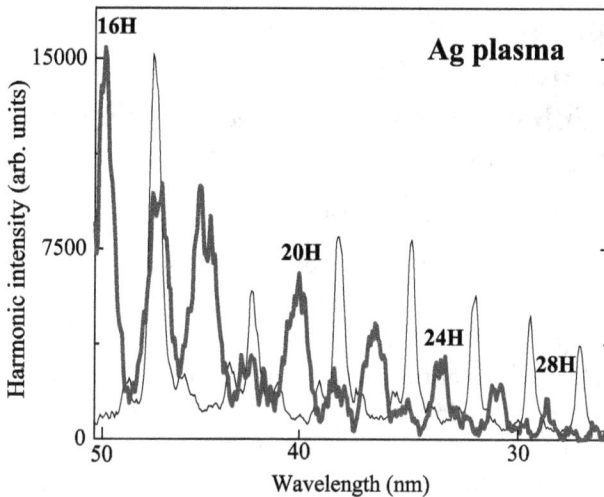

Fig. 6.2 Harmonic spectra for the two-color pump (thick curve) and single-color pump (thin curve) schemes of harmonic generation in silver plasma. One can see that, at a certain phase and intensity relation between the pump waves, the high-order odd harmonics considerably decrease in intensity compared to the even harmonics. Reprinted from [9] with permission from American Physical Society.

relative phase between the two pump waves. At some relative phase between two pumps, one can obtain the harmonic spectrum mostly containing even harmonics, with considerable suppression of odd ones, especially in the shorter wavelength range (thick curve). Note the much larger bandwidth of the even harmonics. Analogous samples of enhanced HHG in the case of the two-color pump can be found in previous chapters.

As underlined in [10], strong harmonic generation is possible due to the formation of a quasilinear field, the selection of a short quantum path component, which has a denser electron wavepacket, and a high ionization rate. The longer time of flight leads to more divergent harmonics in the two-color pump case, taking into consideration only the short trajectories, because the long trajectories will not show up in the measurements in a stable way. With suitable control of the relative phase between the fundamental and second-harmonic field, this particular field significantly enhances the short-path contribution while diminishing other electron paths, resulting in a clean high harmonic spectrum.

6.2. Application of Time-Resolved Spectroscopy of Laser Plasma for Enhancement of Harmonic Efficiency and Generation of Second Plateau in Harmonic Distribution

Analysis of the temporal evolution of spectra for diagnosing laser plasma provides important information about the plasma parameters and can be used for multiple applications. Time-resolved laser-induced plasma spectrometry (TRLIPS) allows the identification of emissions from native molecular species and those due to recombination with ambient air through their different time evolution behavior [11]. The absorption of laser energy by the plasma during its early stage expansion critically influences the properties of the plasma. TRLIPS was also used to analyze the mesh-initiated air breakdown plasma induced by a transverse excitation atmospheric CO_2 pulsed laser [12].

TRLIPS may serve as an important method to study the optimum conditions of plasma formation for achieving efficient HHG of the laser radiation propagating through plasma plumes [13]. It may be noted that most plasma HHG studies were carried out using time-integrated methods of plasma emission analysis, so it was impossible to define exactly what plasma conditions existed during the propagation of the femtosecond pulse through the plume. Since the expansion time of a moderately excited plasma is of the order of tens to hundreds of nanoseconds, it is necessary to temporally synchronize the propagation of the femtosecond pulse through the optimally prepared plasma. It is also important to focus the femtosecond beam in the proper region of the plasma. Most plasma HHG studies were carried out at 20–100 ns delay between the heating picosecond pump pulse and the driving femtosecond pulse. Further increase of the delay leads to a decrease of harmonic efficiency, which emphasizes the importance of the analysis of the dynamics of plasma emission during this period of plasma formation for HHG applications. Below, we present an analysis of time-resolved spectral studies of plasma emission from various metal targets [14]. Those studies were aimed at defining the optimal plasma conditions for efficient HHG in laser plumes and showed that, while over-excitation during laser ablation leads to a drastic decrease of harmonic generation efficiency for most studied plasmas, in some cases one can achieve the conditions for extension of harmonic cutoffs. TRLIPS allows the identification of such plasma plumes.

The time-resolved spectral studies of atomic/ionic emission from laser-produced plasmas were carried out in the UV range (250–300 nm). This spectral range was chosen due to the existence of multiple ionic and atomic transitions in the plasma species used. To create ablation, a 793 nm, 210 ps uncompressed pulse from a Ti:sapphire laser was focused onto metal targets (gallium, platinum, manganese, vanadium, chromium, cadmium, silver, indium) in a vacuum chamber. The second (femtosecond) pulse was propagated through the plasma from the direction orthogonal to the heating beam after some delay with regard to the heating pulse to further excite the plasma plume. The UV spectrum of plasma plumes was observed using a spectrometer and recorded by a time-resolved CCD camera. The time gate for each spectral measurement was 20 ns.

Time-resolved studies of gallium plasma were carried out under both optimum and nonoptimum conditions of plasma formation. These terms refer to maximum and small conversion efficiencies of the femtosecond pulse towards the higher-order harmonics in these plasmas. A change of the gallium plasma emission spectrum was observed in the range of UV spectral lines from neutral atoms and singly charged ions (275–300 nm). The line widths of the four most notable Ga I and Ga II lines at the early stage of plasma formation were approximately equal to the final spectral line widths. At the initial stage of plasma formation, the profile of these spectral lines is governed by the collisions of the emitting atoms with ions and electrons. With increases in delay time, the plasma could be regarded as an equilibrium state in which the probability of collision becomes very small. At gallium plasma concentration of $\sim 5 \times 10^{16}$ cm^{-3}, the collisions do not play an important role, thus leading to the equality of the spectral widths during plasma emission.

The gallium plasma spectra were collected without the delayed excitation by a femtosecond pulse. The influence of a femtosecond pulse on the dynamics of plasma emission was analyzed in the case of ablation of a platinum target. The femtosecond pulse considerably increased the emission from ions, while no noticeable increase in the intensity of neutral lines was observed at the same conditions. Another pattern of the UV spectral dynamics was obtained when the heating pulse intensity was slightly increased at the surface of the platinum target. This increase (from 1.5×10^{10} to 2.2×10^{10} W cm^{-2}) led to a noticeable growth in the emission from the platinum plume. Apart from a considerable growth of the intensities of singly charged ionic lines, some additional lines appeared that showed longer decay times. At these conditions, irradiation by the femtosecond pulse 100 ns after the beginning of laser ablation did not result in a further increase in the intensity of the neutral and ionic lines.

Similar features were observed in the case of vanadium (Fig. 6.3a) and manganese (Fig. 6.3b) plasmas. The manganese plasma spectrum was measured in a narrow UV range (between 254.5 and 257 nm), in which one can compare the excitation dynamics of singly charged manganese ions. The vanadium plasma emission was analyzed in a broader range than other samples. Emission from the V III, V II, and V I ionic and atomic transitions was observed at the optimal conditions. In particular, emission from V III transitions at the

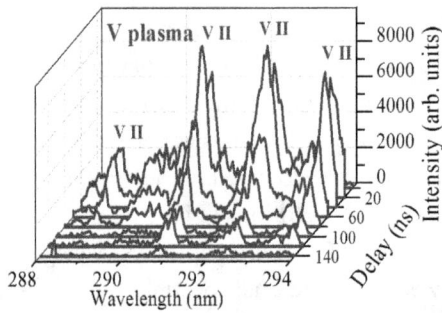

(a)

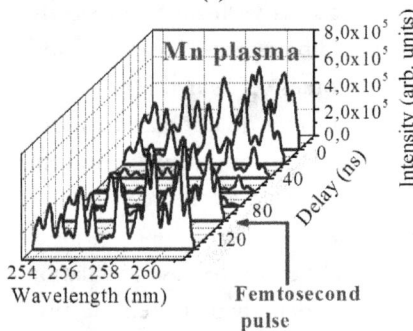

(b)

Fig. 6.3 Time-resolved spectra of (a) vanadium and (b) manganese plasmas. All spectral lines from manganese plasma are identified as arising from Mn II excitation. Reprinted from [4] with permission from American Institute of Physics.

wavelengths of 367.99, 370.53, and 471.49 nm was obtained, and therefore the presence of doubly charged vanadium ions in the laser-ablation plume was confirmed.

Analysis of TRLIPS of various plasma plumes allowed the range of heating pulse and femtosecond pulse intensities at which the harmonic generation was maximal to be defined. This technique provides an efficient method for optimizing the plasma conditions for harmonic generation. Previous studies of HHG from plasma plumes have analyzed the UV emission spectrum from the plasma, aiming to prove that over-excited and/or over-ionized plasma could dramatically decrease the harmonic intensity [15–17]. However, as mentioned above, these spectral measurements were performed using time-integrated methods that did not allow identification of the plasma state before the interaction with the delayed femtosecond pulse. Below, application of TRLIPS for some plasma samples allowing the extension of harmonic cutoffs is presented.

At the initial stages of these studies, when moderate intensities of heating pulse radiation ($I_{pp} = (0.8 - 2) \times 10^{10}\,\mathrm{W\,cm^{-2}}$) were used for the ablation of vanadium and manganese targets, relatively low harmonic cutoffs were achieved (29th and 31st harmonics for the vanadium and manganese plasmas, respectively; see Fig. 6.4a, curve 1, and Fig. 6.4b, curve 1). The growth of heating pulse intensity above $(2-4) \times 10^{10}\,\mathrm{W\,cm^{-2}}$ in the case of these plasma samples led to considerable changes of the plasma spectra in the visible and UV regions, as well as dramatic variations of harmonic distribution in the XUV region. At higher intensities of the heating picosecond pulse ($I_{pp} = (5-7) \times 10^{10}\,\mathrm{W\,cm^{-2}}$), when the excitation of the laser plume was increased and the plasma consisted mostly of singly and doubly charged particles, a new group of harmonics abruptly emerged and exceeded those observed at moderate excitation intensities. This group of harmonics started with one possessing considerably higher intensity than the neighboring lower-order harmonics of the initial (first) plateau. The extension of harmonics depended on the third ionization potentials of the atoms, or the ionization potentials of doubly charged ions. Figures 6.4a (curve 2) and 6.4b (curve 2) show the harmonic spectra observed from the vanadium and manganese plasmas at high excitation of the laser plasma. In these cases, after a steep drop in harmonic intensity near the first cutoffs, the intensity again abruptly increased for the next group of harmonics. One can see a considerable extension of

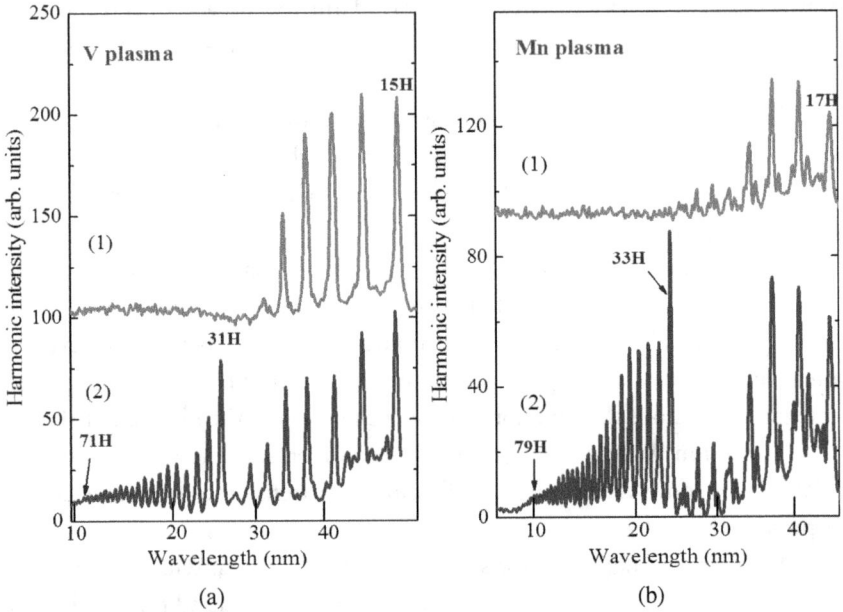

Fig. 6.4 Harmonic spectra from (a) vanadium and (b) manganese plasmas obtained after optimization of plasma excitation using the TRLIPS technique. (1) Optimization for the first plateau distribution. (2) Optimization for the second plateau distribution. Reprinted from [4] with permission from American Institute of Physics.

harmonic cutoffs for these species (71st and 79th harmonics for the vanadium and manganese plumes, respectively). Further optimization of harmonic generation at these conditions by appropriate focusing of the femtosecond pulse in front of plasma plumes allowed additional extension of the harmonic cutoffs in these media (correspondingly up to the 79th (vanadium) [18] and 95th (manganese) harmonics). The characteristic peculiarity of these harmonic spectra was a clearly observable extension of harmonics above previous cutoffs and appearance of the second plateau. This extension can be explained by assuming the involvement of doubly charged ions in HHG. One can note that, at these conditions, the TRLIPS showed spectral lines from both singly and doubly charged ions after 100 ns delay from the beginning of target excitation. The extended harmonic spectra with specific second plateaus were observed also in cadmium and chromium plasmas [19–22]. The maximum harmonic order reported so far from plasma HHG studies relates to such optimization of plasma formation in manganese plumes, with the 101st harmonic generation reported in [23].

Other features of harmonic dynamics were observed in the cases of silver, gallium, indium, and platinum plasma plumes. The increase of heating pulse intensity from 1×10^{10} to $4 \times 10^{10}\,\mathrm{W\,cm^{-2}}$ during HHG experiments led to the appearance of strong plasma emission at wavelengths above 20 nm. At these excitation conditions, the harmonic emission from silver plasma was overlapped by the plasma emission. The intensity of generated harmonics became considerably less than at $1 \times 10^{10}\,\mathrm{W\,cm^{-2}}$ excitation. One can note that analogous behavior was observed in molybdenum [24, 25], beryllium [26], magnesium [27], platinum [28], and aluminum [29]. In the case of gallium and indium plasmas, this over-excitation led to the entire disappearance of harmonic emission, while at the optimal excitation defined from TRLIPS analysis, the generation of almost "clean" harmonic spectra was obtained.

The reasons why the conditions for extended harmonic generation were achieved for a few targets, while in most other cases there was a drop of HHG conversion efficiency, are not clear. It may be that the role of additional free electrons appearing during the ionization of singly charged ions in the cases of manganese, chromium, cadmium, and vanadium can be less influential compared with other plasma plumes, due to the influence of transitions possessing strong oscillator strengths in the spectral ranges of the initial cutoffs. Indeed, the reported transitions of chromium ions in the region above 40 eV showed high values of oscillator strengths. Previous studies of the photoabsorption and photoionization spectra of chromium plasma in the range of 41–42 eV have demonstrated the presence of strong transitions [30–32]. In particular, the region of the "giant" 3p–3d resonance of Cr II spectra was analyzed in [32], and strong transitions, which could enhance the optical and nonlinear optical response of the plume, were revealed. Studies of the photoionization spectra of cadmium ions have also shown the presence of analogous strong ionic transitions [33]. The above observations and assumptions allows us to assume a decisive role for the spectral characteristics of ions in the formation of the optimal conditions when the negative influence of the free electrons on HHG conversion efficiency can be compensated for by the closeness of the next set of harmonics to the ionic transitions possessing strong oscillator strengths.

6.3. Application of Carbon Aerogel Plumes as Efficient Media for HHG in the 40–90 nm Range

In Chapter 5, we showed the advanced properties of carbon-containing plasmas produced on the surfaces of graphite, carbon nanotubes, and C_{60}. Among the carbon-containing plasma media that could show strong HHG conversion efficiency, one can also consider carbon aerogels. Carbon aerogels are a relatively new class of highly porous, nanostructured, open-cell monolithic materials and have received much current interest due to their unique thermal, optical, and electrical properties, most of which are attributed to special structural features [34]. Carbon aerogels exhibit high electrical conductivity and high open porosity (>50%). These properties are of crucial importance when one considers the efficient absorption of laser radiation during the irradiation of an ablating surface.

In [35], the preparation and characterization of carbon aerogels, as well as the study of HHG in carbon aerogel plasma plumes, was carried out. A higher yield of HHG was observed primarily for the lower-order harmonics in the wavelength range of 40–90 nm. The harmonic yield was two times higher than for the C_{60} plumes for odd harmonics using a single-color laser. Next, in the case of a two-color pump, the yield was observed to be comparable for both odd and even harmonics. The observed conversion efficiency from the carbon aerogel plasma was shown to be the highest compared to earlier studies in metal plasma plumes, and it slightly exceeded the harmonic efficiency from the fullerene-containing plasma, which has been reported as one of the most effective media for plasma HHG [36].

Figure 6.5 shows the HHG spectra obtained from carbon aerogel plasma using both single-color and two-color laser beams. The thick curve shows the spectra using both the 400 nm and 800 nm beams, and the thin curve was obtained using a single-color laser beam. Both the odd and even harmonics showed equal intensities and the harmonic cutoff falls rapidly at the 19th to 21st orders, i.e., in the region of 40 nm. Note that the harmonic cutoffs in the cases of single-color and two-color pumps were not much different. The harmonic cutoffs from the carbon aerogel plasma in the case of the single-color pump were observed in the range between the 23rd and 29th harmonics, and they did not strongly depend on the conditions of the experiments.

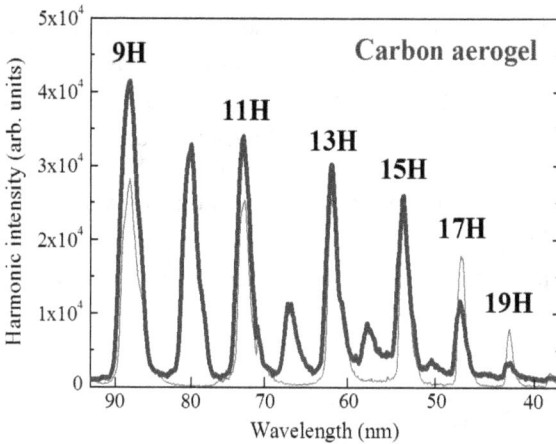

Fig. 6.5 Comparison of the HHG spectra of carbon aerogel plasma in the cases of the single-color pump (thin curves) and the two-color pump (thick curves). Reprinted from [35] with permission from Optical Society of America.

The reported conversion efficiency from fullerene plasma (5×10^{-5} [37–39]) can also be attributed to the carbon aerogel plasma, because the experiments analyzed in this work have demonstrated the similarity in the HHG conversion efficiencies in these two plasma plumes.

The important issue of these and other [40–43] studies is why carbon-containing plasma generates intense harmonics. We presented in this and previous chapters some observations of this phenomenon. Assumptions that could explain the high harmonic yield from carbon-containing media, and in particular aerogel plasma, are as follows:

1. Carbon targets allow for easier generation of relatively dense plasma and corresponding phase matching conditions for lower-order harmonics.
2. The first ionization potential of carbon is high enough to prevent the appearance of a high concentration of free electrons compared with other plasmas.
3. Neutral carbon atoms dominate in the carbon vapor before the interaction with the short laser pulse.
4. Phase mismatch due to free electrons is small, and the effective laser intensity that the medium can experience is higher, because laser radiation defocusing effects are negligible.

5. The porosity of the carbon aerogel can create the conditions for efficient transformation from a solid to a gaseous phase during ablation by picosecond pulses. This allows for the creation of optimal conditions when plasma properties satisfy the most efficient harmonic generation.

Thus, the specific plasma formation conditions during laser ablation of carbon-containing plasmas prepared, in particular, on the surface of carbon aerogel can cause efficient frequency conversion of femtosecond pulses in this medium.

6.4. Comparative Studies of HHG in Laser Plasmas and Gases

There have been various efforts to create more efficient HHG sources, including techniques using gas and plasma media, the use of cluster media that enhance harmonic intensity [44–46], use of modulated hollow-core waveguides for adjusting phase matching [47], quasi-phase matching with several gas jets [48], and resonance-induced enhancement [49]. These investigations led to generation of microjoule-level harmonic pulses both in gases [50] and in carbon plasmas [40, 41]. The promising earlier results from carbon laser plasma plumes comprising fullerenes [38], carbon nanotubes [43], and graphite [40, 41] for HHG in the 40–100 nm range, where considerable enhancements in harmonic yield were observed, prompt further investigation. These observations motivated a comparison of the efficiencies of HHG in carbon-containing plasmas and gases under closely matched experimental conditions. Below, we discuss studies that were carried out using graphite plasmas and argon gas. The measurements show that, in the 14–28 eV photon energy range, carbon plumes are more efficient than argon. The likely reason for this enhancement is the high nonlinear response of carbon nanoparticles, which were observed in the experiments, though their involvement in HHG requires further studies [51].

In the experiments, two different lasers were employed to create the carbon ablation plume. In the first case, 10% (120 μJ) of the uncompressed pulse energy of a 1 kHz Ti:sapphire laser, with central wavelength 800 nm and pulse duration 8 ps, was split from the beam line prior to the laser compressor stage and was focused by a 400 mm focal length lens into the vacuum chamber to create a plasma on the targets using an intensity on the target surface of typically $I_{pp} = 2 \times 10^{10} \, \text{W cm}^{-2}$. In the second case, a separate Nd:YAG laser (1064 nm, 7 mJ, 10 ns, 10 Hz) was used for plasma formation, which created plasma plumes at lower intensity but at higher fluence on the target surface. The Nd:YAG beam was focused by a 400 mm focal length lens and created a plasma plume with a diameter of ~0.35 mm using an intensity on the target surface of $I_{np} = 1 \times 10^9 \, \text{W cm}^{-2}$.

The Ti:sapphire laser after compression provided pulses of 25 fs duration and energies of up to 0.8 mJ at a repetition rate of 1 kHz. The amplified pulses were focused into a 1 m long differentially pumped hollow-core fiber filled with neon (250 μm inner core diameter) [52]. The spectrally broadened pulses at the output of the fiber system were compressed by ten bounces off double-angle technology chirped mirrors. In this way, high-intensity few-cycle pulses (0.2 mJ, $\leq$5 fs) [53] were generated.

These few-cycle pulses were focused into the plasma plume, approximately 150–200 μm above the target surface, to generate high-order harmonics using a 400 mm focal length reflective mirror. The delay between plasma initiation and the arrival at the plume of the few-cycle pulse when using 10 ns ablation pulses was varied in the range of 10–120 ns by varying the Nd:YAG laser Q-switch delay. For the 10 ns ablation pulses, the studies showed that delays between 30 and 40 ns are optimal to achieve the highest conversion efficiency and these were chosen for further improvement of harmonic yield. In the case of plasma formation by 8 ps pulses the delay was fixed at 34 ns. The plasma plume and target were moved over a range of positions with respect to the focal point of the few-cycle pulse, to optimize the conversion efficiency of harmonic generation. The laser intensity in the plume was estimated to be $I_{fp} = 5 \times 10^{14} \, \text{W cm}^{-2}$. The HHG radiation was analyzed by a spatially resolving XUV spectrometer consisting of a flat-field grating and an imaging microchannel plate detector read out by a CCD camera. The gas HHG was performed in a tubular target (length ~1.5 mm, diameter 0.5 mm) filled with argon or neon by a piezo-actuated pulsed valve operating at 1 kHz.

The application of the shortest available pulses for studies of HHG in laser plumes is crucial for further enhancement of harmonic yield. A short laser pulse duration minimizes ionization of atoms, molecules, and clusters, as well as reducing the fragmentation of molecules and clusters. Given that nanoparticles of low ionization potential are likely to be present in the plumes, short laser pulses are critically important to enable the observation of enhanced harmonic generation from nanoparticles. Three types of targets (bulk graphite, boron carbide (B_4C), and carbon aerogel plates) for plasma harmonics and two gases (argon and neon) for gas harmonic generation were investigated in these studies. Harmonics from carbon-containing plasmas were optimized by choosing the delay between pulses, distance between the femtosecond beam and target, and z-position of the plasma (where z is the HHG drive beam propagation axis) at which maximum harmonic efficiency was obtained.

As the femtosecond laser intensity increased, the growth of harmonic intensity, extension of cutoff, and blueshift of harmonics from the 10 ns pulse-induced plasma were observed. A smaller blueshift of higher harmonics from gas targets, which are not strongly ionized for experimental conditions, was observed. No blueshift of harmonics was observed in the case of plasmas produced by 8 ps pulses, due to the lower density of the plasma in that case.

The studies of harmonic generation efficiencies in plasmas and gases were carried out at identical laser conditions for the few-cycle laser. The conversion efficiency in both media was optimized and the relative intensities of harmonics were measured in different spectral ranges. Figure 6.6a shows typical harmonic spectra for the 9th, 11th, and 13th harmonic orders from the carbon plasma (graphite target) and gas (argon) at these conditions. Argon is the most widely used gas for HHG, representing a good trade-off between efficiency and cost. These studies were performed using the 10 ns pump pulses for creation of the plasma plume. It was found that for the 9th to 15th harmonics the conversion efficiency from the carbon plasma was two to five times greater than from the gas. Higher orders (>15) were more efficiently generated in the gas than in the plasma. The same comparison was made when the plasma was created by the 8 ps pulses. In that case, the conversion efficiency for the same harmonics (9–15) in argon was two to five times higher with respect to the carbon plasma harmonics. The highest harmonic order observed from the carbon plasma was 27 (Fig. 6.6b). One can note that an efficient ultrafast radiation source in the spectral range 14–25 eV, which corresponds to the 9th to 15th harmonics of a Ti:sapphire laser, is extremely useful because it allows the investigation of valence electron dynamics in many important chemical and biological systems.

To explain the observation of enhanced HHG in the case of the 10 ns pulse-induced carbon plasma, the appearance of nanoparticles in the plasma

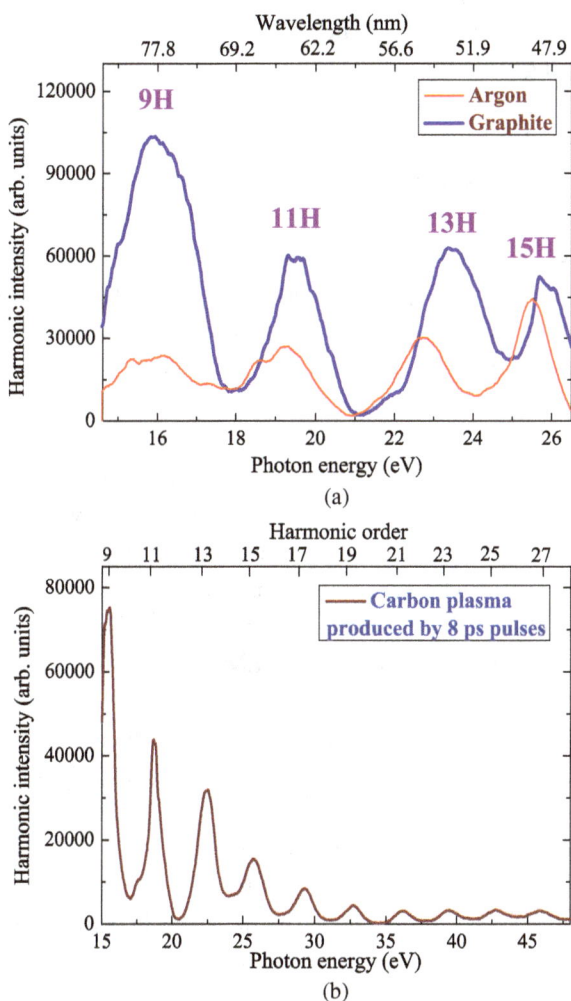

Fig. 6.6 (a) Low-order harmonic spectra from carbon plasma produced by 10 ns pulses and argon gas at the optimal conditions of harmonic generation for both media. (b) Harmonic spectrum from carbon plasma in the case of plasma formation using 8 ps pulses. Reprinted from [51] with permission from American Physical Society.

during laser ablation, which can enhance the harmonic yield, has been implicated [54]. The nanoparticles can be formed on the ablated surface and transferred to the plasma where they enhance the efficiency of HHG. A likely explanation for more efficient harmonic generation from nanoparticles compared with single atoms/ions is the higher local concentration of neutral atoms in the nanoparticles and the enhanced cross section of recombination for the returning electron.

To prove the presence of nanoparticles in ablation plasmas, the morphology of deposited debris from the graphite plasma during ablation using the picosecond and nanosecond pulses was analyzed. The debris was collected on a glass surface placed 40 mm from the ablated target. Laser ablation is a widely accepted technique for forming nanoparticles [55, 56]. However, this process has previously been optimized without consideration of the formation of free electrons and highly excited ions, which destroy the optimal conditions for phase matched HHG. One has to emphasize that morphology measurements were carried out at the laser ablation conditions corresponding to the maximum HHG conversion efficiency observed. At these conditions, the SEM images did not show the presence of nanoparticles in the deposited debris from the 8 ps pulse-induced plasma when the highest conversion efficiency of harmonics from carbon-containing plasma was obtained. Another interesting feature was observed when the target was ablated using the 10 ns pulses. At a relatively moderate ablation intensity ($\sim 1 \times 10^9 \, \mathrm{W \, cm^{-2}}$), an abundance of nanoparticles was observed in the SEM images of the collected debris, with the sizes mostly distributed in the range between 20 and 100 nm. This difference in the morphology of deposition is probably due to the considerably lower fluence on the target of the picosecond pulses ($0.2 \, \mathrm{J \, cm^{-2}}$) compared to the nanosecond pulses ($10 \, \mathrm{J \, cm^{-2}}$). Thus, the morphological studies have confirmed the presence of significant quantities of nanoparticles in the carbon plasma produced by 10 ns pulses. It was also found that the efficiency of low-order harmonics from the carbon plasma produced by 8 ps pulses was stronger with respect to some other noncarbon plasma plumes, such as aluminum, silver, and chromium plasmas, which was confirmed in a separate set of measurements. It is possible that in the case of carbon plasmas produced at these conditions, harmonics could also originate from the nanoparticles. However, their sizes may be so small (<2–5 nm) that the resolution of SEM images becomes insufficient for their identification.

The observations of strong harmonic yield from both carbon plasmas at short delay may point to the simultaneous appearance of nanoparticles together with atoms and ions in the area of femtosecond pulse propagation. The delay between the ablation forming and HHG drive pulses was varied over a broad range (from 0 to a few microseconds) and harmonics observed only in the 6–180 ns delay range. This indicates that both single particles and nanoparticles arrive in the femtosecond pulse interaction volume at the same time (i.e., 30–40 ns from the onset of ablation).

To avoid the uncertainties of absolute harmonic flux measurements encountered in earlier work [57], a direct comparison of gas and plasma harmonics under the same conditions was carried out. The studies demonstrated a higher conversion efficiency for the low-order harmonics generated in graphite plasma compared with argon gas in the case of 10 ns ablating pulses. The density of the argon gas jet used in the comparative studies was measured by the absorption of XUV harmonics. Harmonics in the range of 30–60 eV were generated using a tube target filled with neon and propagated through the argon gas jet that was placed 5 cm after the focus. The difference in harmonic intensity with and without argon gas in the jet was recorded. Using known values of the argon absorption cross section in this energy range [58] and assuming a uniform gas distribution of 1.5 mm in length, an argon density of $\sim 6 \times 10^{17}$ cm^{-3} was estimated for the same operating conditions as used in the comparative study.

The absorption measurements were carried out at the conditions of plasma ablation when the maximum HHG conversion efficiency was maintained and did not register the attenuation of harmonics propagating through the plasma due to the small absorption cross section of carbon in the range of 14–30 eV. Instead, a three-dimensional molecular dynamical simulation of laser ablation of graphite was performed using the open-source molecular dynamics code ITAP IMD [59, 60]. One can note that molecular dynamical simulations are reliable and widely used [61]. They can also treat cluster formation and are able to handle short-duration ablation pulses and different focusing geometries for the ablation beam. The modeling was performed under conditions in which ablated atoms gain no further kinetic energy from the ablation pulse.

The simulation was performed for both ablation pulse durations (8 ps and 10 ns) with a time step of 1 fs. The positions and velocities of all carbon atoms were fixed every 50 fs. The velocity distribution was analyzed by counting the number of particles that have sufficient velocity to be in a given region (200 μm above the ablated target) 30 ns after the end of the pulse. Then the number of particles was averaged for all measurements and divided by the volume. This volume is taken as the surface of the model sample multiplied by time and by the difference between minimal and maximal speeds allowed. The density of the carbon plasma was calculated at the experimental conditions of target ablation (i.e., 2×10^{10} W cm^{-2} in the case of 8 ps pulses and 1×10^9 W cm^{-2} in the case of 10 ns pulses). The corresponding densities were found to be 2.6×10^{17} cm^{-3} and 2.5×10^{18} cm^{-3}.

One can note that there was no significant blueshift of harmonics in the case of HHG in picosecond pulse-induced carbon plasma, while in the case of nanosecond pulse-induced plasma, a clear blueshift was observed for all harmonics. This confirms that the concentration of particles in the case of picosecond pulse-induced ablation was less than that for nanosecond pulse-induced carbon plasma. One can make a simple estimate of the nonlinear response of a single particle from carbon plasma relative to an argon atom by writing the intensity of the qth-order harmonic as [61]

$$I_q \propto |d_q|^2 N^2 |L^2| \qquad (6.1)$$

where d_q is the dipole moment of the emitter at frequency $q\omega$, $|d_q|^2$ denotes the single particle nonlinear optical response, N is the emitter density, and L is the length of the medium. For the experiment with 10 ns heating pulses, $L^{(Ar)} = 1.5$ mm for the argon gas jet, and $L^{(plasma)} = 0.35$ mm for carbon plasma produced by 10 ns pulses. It follows from Eq. 6.1 that

$$\frac{|d_q^{(plasma)}|^2}{|d_q^{(Ar)}|^2} \approx \frac{I_q^{(plasma)}}{I_q^{(Ar)}} \left(\frac{N^{(Ar)} L^{(Ar)}}{N^{(plasma)} L^{(plasma)}} \right)^2 \qquad (6.2)$$

Using the measured ratio $I_q^{(plasma)} / I_q^{(Ar)} \approx 2-5$ and for $N^{(Ar)} / N^{(plasma)} \approx 0.24$, and $L^{(plasma)} / L^{(Ar)} \approx 4$ one can estimate that $|d_q^{(plasma)}|^2 / |d_q^{(Ar)}|^2 \approx 2-5$ in the case of ablation using the 10 ns pulses. The harmonic emitters in the plasma may be atoms and ions, as well as nanoparticles of different sizes. Hence, this estimate refers to the average value of the nonlinear optical response based on the average densities of the plasma media. A likely explanation for the increased nonlinear response of the ablation media is the presence of carbon nanoparticles detected in the plume. These clusters can increase the cross section of recombination for the accelerated electron, thus enhancing the yield of harmonics.

Analogous estimates in the case of ablation by 8 ps pulses ($N^{(plasma)} = 2.6 \times 10^{17}$ cm^{-3}, $L^{(plasma)} = 0.25$ mm, $I_q^{(plasma)} / I_q^{(Ar)} \approx 0.2-0.5$) showed

Table 6.1. Comparison of experimental parameters during plasma and gas HHG.

Medium	10 ns pulse-produced carbon plasma	8 ps pulse-produced carbon plasma	Argon gas
HHG drive intensity in the medium	$4 \times 10^{14} \, \mathrm{W \, cm^{-2}}$	$4 \times 10^{14} \, \mathrm{W \, cm^{-2}}$	$4 \times 10^{14} \, \mathrm{W \, cm^{-2}}$
Ablation laser intensity on the target surface	$1 \times 10^{9} \, \mathrm{W \, cm^{-2}}$	$2 \times 10^{10} \, \mathrm{W \, cm^{-2}}$	
Ablation laser fluence on the surface	$5 \, \mathrm{J \, cm^{-2}}$	$0.2 \, \mathrm{J \, cm^{-2}}$	
Length of nonlinear medium	0.35 mm	0.25 mm	1.5 mm
Density of the medium	$2.5 \times 10^{18} \, \mathrm{cm^{-3}}$	$2.6 \times 10^{17} \, \mathrm{cm^{-3}}$	$6 \times 10^{17} \, \mathrm{cm^{-3}}$
Relative intensities of 9th to 13th harmonics	2–5	0.2–0.5	1
Relative density–length product	1	13.8	1
Relative averaged nonlinear optical response	2–5	40–100	1

that the parameter $|d_q^{(\mathrm{plasma})}|^2 / |d_q^{(\mathrm{Ar})}|^2$ is higher by more than one order of magnitude compared with the case of plasma formation using 10 ns pulses ($|d_q^{(\mathrm{plasma})}|^2 / |d_q^{(\mathrm{Ar})}|^2 \approx 40-100$). Again, as in the case of plasma produced by 10 ns pulses, for 8 ps pulse-produced plasma, this enhancement could be due to the presence of small nanoparticles ($<5 \, \mathrm{nm}$), which can enhance the yield of harmonic emission. The parameters for comparison between the three targets are summarized in Table 6.1.

In addition to graphite targets, other carbon-containing materials were studied in an effort to obtain the best plasma for achieving efficient harmonics with stable output characteristics. Three carbon-containing targets were compared, which can be distinguished by their different hardness. Two targets — bulk B_4C and relatively soft carbon aerogel — were used for ablation by 8 ps pulses and compared with the efficiency and stability of harmonic yield from the graphite plasma. High harmonics were produced with the highest efficiency in the plasma from the B_4C target.

Another important advantage with the B_4C target is the improved shot-to-shot stability of the harmonic intensity at kilohertz pulse repetition rates

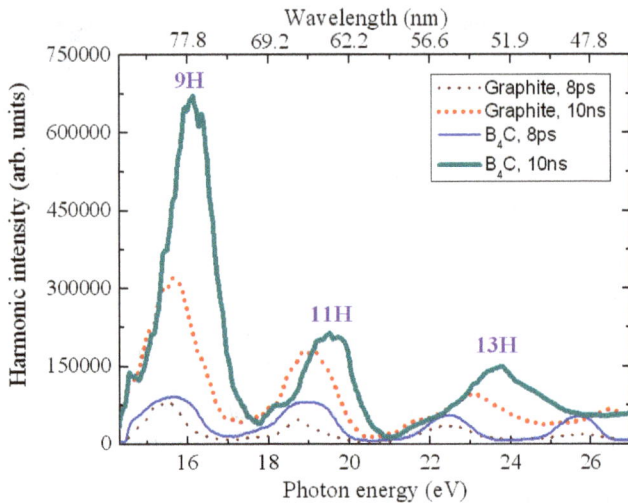

Fig. 6.7 Comparison of low-order harmonic spectra from B_4C (solid lines) and graphite (dotted lines) plasmas in the cases of 10 ns (thick lines) and 8 ps (thin lines) ablating pluses. Reprinted from [51] with permission from American Physical Society.

compared to the softer targets. Thus B_4C could be a good candidate for applications of plasma harmonics including pulse duration measurements. These advantages were confirmed during comparative studies of harmonics from graphite and B_4C plasmas in the case of target ablation using 10 ns pulses (Fig. 6.7). One can see that harmonics from the B_4C plasma are stronger than those from the graphite plasma also for longer pulses.

References

1. Chen, G., Chen, J.G., Yang, Y.J. *et al.* (2010). Effect of the relative phase between two-colour pump pulses on structure of harmonic spectra, *Eur. Phys. J. D*, **57**, 145–149.
2. Cormier, E. and Lewenstein, M. (2000). Optimizing the efficiency in high order harmonic generation optimization by two-colour fields, *Eur. Phys. J. D*, **12**, 227–233.
3. Kim, I.J., Kim, C.M., Kim, H.T. *et al.* (2005). Highly efficient high-harmonic generation in an orthogonally polarized two-colour laser field, *Phys. Rev. Lett.*, **94**, 243901.
4. Mauritsson, J., Johnsson, P., Gustafsson, E. *et al.* (2006). Attosecond pulse trains generated using two colour laser fields, *Phys. Rev. Lett.*, **97**, 013001.
5. Pfeifer, T., Gallmann, L. and Abel, M.J. (2006). Single attosecond pulse generation in the multicycle-driver regime by adding a weak second-harmonic field, *Opt. Lett.*, **31**, 975–977.
6. Charalambidis, D., Tzallas, P., Benis, E.P. *et al.* (2008). Exploring intense attosecond pulses, *New J. Phys.*, **10**, 025018.

7. Kim, I.J., Lee, G.H., Park, S.B. *et al.* (2008). Generation of submicrojoule high harmonics using a long gas jet in a two-color laser field, *Appl. Phys. Lett.*, **92**, 021125.

8. Ganeev, R.A., Singhal, H., Naik, P.A. *et al.* (2009). Enhancement of high-order harmonic generation using two-colour pump in plasma plumes, *Phys. Rev. A*, **80**, 033845.

9. Ganeev, R.A., Singhal, H., Naik, P.A. *et al.* (2010). Systematic studies of two-colour pump induced high order harmonic generation in plasma plumes, *Phys. Rev. A*, **82**, 053831.

10. Pont, M. (1989). Atomic distortion and ac-Stark shifts of H under extreme radiation conditions, *Phys. Rev. A*, **40**, 5659–5672.

11. Boueri, M., Baudelet, M., Yu, J. *et al.* (2009). Early stage expansion and time-resolved spectral emission of laser-induced plasma from polymer, *Appl. Surf. Sci.*, **255**, 9566–9571.

12. Camacho, J.J., Diaz, L., Santos, M. *et al.* (2010). Time-resolved optical emission spectroscopy of laser-produced air plasma, *J. Appl. Phys.*, **107**, 083306.

13. Elouga Bom, L.B., Kieffer, J.-C., Ganeev, R.A. *et al.* (2007). Influence of the main pulse and prepulse intensity on high-order harmonic generation in silver plasma ablation, *Phys. Rev. A*, **75**, 033804.

14. Ganeev, R.A., Elouga Bom, L.B. and Ozaki, T. (2011). Time-resolved spectroscopy of plasma plumes: A versatile approach for optimization of high-order harmonic generation in laser plasma, *Phys. Plasmas*, **18**, 083101.

15. Ganeev, R.A., Singhal, H., Naik, P.A. *et al.* (2006). Harmonic generation from indium-rich plasmas, *Phys. Rev. A*, **74**, 063824.

16. Ganeev, R.A., Baba, M., Suzuki, M. *et al.* (2006). Optimization of harmonic generation from boron plasma, *J. Appl. Phys.*, **99**, 103303.

17. Ganeev, R.A., Suzuki, M., Baba, M. *et al.* (2007). High harmonic generation from the laser plasma produced by the pulses of different duration, *Phys. Rev. A*, **76**, 023805 (2007).

18. Suzuki, M., Baba, M., Kuroda, H. *et al.* (2007). Seventy first harmonic generation from ions in laser ablation vanadium plume at 11.2 nm, *Opt. Express*, **15**, 4112–4117.

19. Ganeev, R.A., Suzuki, M., Baba, M. *et al.* (2009). Extended high-order harmonic spectra from the laser-produced Cd and Cr plasmas, *Appl. Phys. Lett.*, **94**, 051101.

20. Ganeev, R.A., Suzuki, M., Baba, M. *et al.* (2005). Harmonic generation in XUV from chromium plasma, *Appl. Phys. Lett.*, **86**, 131116.

21. Ganeev, R.A., Elouga Bom, L.B., Kieffer, J.-C. *et al.* (2007). Systematic investigation of resonance-induced single harmonic enhancement in the extreme ultraviolet range, *Phys. Rev. A*, **75**, 063806.

22. Suzuki, M., Ganeev, R.A., Elouga Bom, L.B. *et al.* (2007). Extension of cutoff in high-harmonic by using doubly charged ions in a laser-ablation plume, *J. Opt. Soc. Am. B*, **24**, 2847–2852.

23. Ganeev, R.A., Elouga Bom, L.B., Kieffer, J.-C. *et al.* (2007). Demonstration of the 101st harmonic generated from laser-produced manganese plasma, *Phys. Rev. A*, **76**, 023831.

24. Ganeev, R.A., Kulagin, I.A., Suzuki, M. *et al.* (2005). Harmonic generation in Mo plasma, *Opt. Commun.*, **249**, 569–577.

25. Ganeev, R.A., Suzuki, M., Baba, M. *et al.* (2005). High-order harmonic generation from carbon plasma, *J. Opt. Soc. Am. B*, **22**, 1927–1933.

26. Ganeev, R.A., Suzuki, M., Baba, M. *et al.* (2008). Application of beryllium plasma for the high-order harmonic generation of Ti:sapphire laser radiation, *J. Opt. Soc. Am. B*, **25**, 2096–2100.

27. Ganeev, R.A. and Kuroda, H. (2005). Frequency conversion of femtosecond radiation in magnesium plasma, *Opt. Commun.*, **256**, 242–247.
28. Ganeev, R.A., Elouga Bom, L.B., Kieffer, J.-C. *et al.* (2007). Optimum plasma conditions for the efficient high-order harmonic generation in platinum plasma, *J. Opt. Soc. Am. B*, **24**, 1319–1323.
29. Ganeev, R.A., Baba, M., Suzuki, M. *et al.* (2006). 33rd harmonic generation from aluminum plasma, *J. Mod. Opt.*, **53**, 1451–1458.
30. West, J.B., Hansen, J.E., Kristensen, B. *et al.* (2003). Revised interpretation of the photoionization of Cr^+ in the 3p excitation region, *J. Phys. B: At. Mol. Opt. Phys.*, **36**, L327–L334.
31. West, J.B. (2001). Photoionization of atomic ions, *J. Phys. B: At. Mol. Opt. Phys.*, **34**, R45–R92.
32. McGuinness, C., Martins, M., Wernet, P. *et al.* (1999). Metastable state contributions to the measured 3p photoabsorption spectrum of Cr^+ ions in a laser-produced plasma, *J. Phys. B: At. Mol. Opt. Phys.*, **32**, L583–L592.
33. Kilbane, D., Kennedy, E.T., Mosnier, J.-P. *et al.* (2005). EUV photoabsorption spectra of Cd II and Cd III, *J. Phys. B: At. Mol. Opt. Phys.*, **38**, 83–88.
34. Horikawa, T., Hayashi, J. and Muroyama, M. (2004). Controllability of pore characteristics of resorcinol–formaldehyde carbon aerogel, *Carbon*, **42**, 1625–1633.
35. Ganeev, R.A., Naik, P.A., Chakera, J. A. *et al.* (2011). Carbon aerogel plumes as an efficient medium for higher harmonic generation in 40–90 nm range, *J. Opt. Soc. Am. B*, **28**, 360–364.
36. Ganeev, R.A. (2011). Fullerenes: the attractive medium for harmonic generation, *Laser Phys.*, **21**, 25–43.
37. Ganeev, R.A., Elouga Bom, L.B., Wong, M.C.H. *et al.* (2009). High-order harmonic generation from C_{60}-rich plasma, *Phys. Rev. A*, **80**, 043808.
38. Ganeev, R.A., Singhal, H., Naik, P.A. *et al.* (2009). Influence of C60 morphology on high-order harmonic generation enhancement in fullerene-containing plasma, *J. Appl. Phys.*, **106**, 103103.
39. Ganeev, R.A., Elouga Bom, L.B., Abdul-Hadi, J. *et al.* (2009). High-order harmonic generation from fullerene using the plasma harmonic method, *Phys. Rev. Lett.*, **102**, 013903.
40. Elouga Bom, L.B., Petrot, Y., Bhardwaj, V.R. *et al.* (2011). Multi-μJ coherent extreme ultraviolet source generated from carbon using the plasma harmonic method, *Opt. Express*, **19**, 3077–3085.
41. Petrot, Y., Elouga Bom, L.B., Bhardwaj, V.R. *et al.* (2011). Pencil lead plasma for generating multimicrojoule high-order harmonics with a broad spectrum, *Appl. Phys. Lett.*, **98**, 101104.
42. Ganeev, R.A., Singhal, H., Naik, P.A. *et al.* (2010). Enhanced harmonic generation in C_{60}-containing plasma plumes, *Appl. Phys. B*, **100**, 581585.
43. Ganeev, R.A., Naik, P.A., Singhal, H. *et al.* (2011). High order harmonic generation in carbon nanotube-containing plasma plumes, *Phys. Rev. A*, **83**, 013820.
44. Donnelly, T.D., Ditmire, T., Neuman, T. *et al.* (1996). High-order harmonic generation in atom clusters, *Phys. Rev. Lett.*, **76**, 2472–2475.
45. Tisch, J.W.G., Ditmire, T., Frasery, D.J. *et al.* (1997). Investigation of high-harmonic generation from xenon atom clusters, *J. Phys. B: At. Mol. Opt. Phys.*, **30**, L709–L713.

46. Singhal, H., Ganeev, R.A., Naik, P.A. *et al.* (2010). Study of high-order harmonic generation from nanoparticles, *J. Phys. B: At. Mol. Opt. Phys.*, **43**, 025603.
47. Ferray, M., L'Huillier, A., Li, X.-F. *et al.* (1988). Multiple-harmonic conversion of 1064 nm radiation in rare gases, *J. Phys. B: At. Mol. Opt. Phys.*, **21**, L31–L36.
48. Pirri, A., Corsi, C. and Bellini, M. (2008). Enhancing the yield of high-order harmonics with an array of gas jets, *Phys. Rev. A*, **78**, 011801.
49. Ganeev, R.A., Suzuki, M., Ozaki, T. *et al.* (2006). Strong resonance enhancement of a single harmonic generated in extreme ultraviolet range, *Opt. Lett.*, **31**, 1699–1701.
50. Hergott, J.-F., Kovacev, M., Merdji, H. *et al.* (2002). Extreme-ultraviolet high-order harmonic pulses in the microjoule range, *Phys. Rev. A*, **66**, 021801.
51. Ganeev, R.A., Witting, T., Hutchison, C. *et al.* (2012). Enhanced high-order harmonic generation in a carbon ablation plume, *Phys. Rev. A*, **85**, 015807.
52. Robinson, J.S., Haworth, C.A., Teng, H. *et al.* (2006). The generation of intense, transform-limited laser pulses with tuneable duration from 6 to 30 fs in a differentially pumped hollow fibre, *Appl. Phys. B*, **85**, 525–529.
53. Witting, T., Frank, F., Arrell, C.A. *et al.* (2011). Characterization of high-intensity sub-4-fs laser pulses using spatially encoded spectral shearing interferometry, *Opt. Lett.*, **36**, 1680–1682.
54. Ganeev, R.A., Suzuki, M., Baba, M. *et al.* (2008). High-order harmonic generation in Ag nanoparticle-contained plasma, *J. Phys B: At. Mol. Opt. Phys.*, **41**, 045603.
55. He, C., Sasaki, T., Shimizu, Y. *et al.* (2008). Synthesis of ZnO nanoparticles using nanosecond pulsed laser ablation in aqueous media and their self-assembly towards spindle-like ZnO aggregates. *Appl. Surf. Sci.*, **254**, 2196–2202.
56. Siems, A., Weber, S.A.L., Boneberg, J. *et al.* (2011). Thermodynamics of nanosecond nanobubble formation at laser-excited metal nanoparticles, *New J. Phys.*, **13**, 043018.
57. Ozaki, T., Elouga Bom, L.B., Abdul-Hadi, J. *et al.* (2010). Blueprint for generating intense attosecond pulses using high-order harmonics from graphitic carbon plasma, *Proceedings 23rd Meeting of the IEEE Photonics Society* (7–11 November 2010, Denver, USA), pp. 684–685.
58. Henke, B.L., Gullikson, E.M. and Davis, J.C. (1993). X-ray interactions: Photoabsorption, scattering, transmission, and reflection at $E = 50-30,000 \, eV$, $Z = 1-92$, *Atom. Data Nucl. Data Tables*, **54**, 181–342.
59. Stadler, J., Mikulla, R. and Trebin, H.-R. (1997). IMD: a software package for molecular dynamics studies on parallel computers, *Int. J. Modern Phys. C*, **8**, 1131–1140.
60. Roth, J., Géahler, F. and Trebin, H.-R. (2000). A molecular dynamics run with 5 180 116 000 particles, *Int. J. Modern Phys. C*, **11**, 317–322.
61. Altucci, C., Bruzzese, R., de Lisio, C. *et al.* (2000). Tuneable soft-x-ray radiation by high-order harmonic generation, *Phys. Rev. A*, **61**, 021801.

7

Recent Developments and Future Perspectives of Plasma HHG

This chapter describes the most recent developments in the field of plasma HHG. Here we present new trends, schemes, and approaches in plasma harmonics, harmonic generation in carbon nanotube-containing plasma plumes, destructive interference of laser harmonics in mixtures of various emitters in plasmas, generation of broadband harmonics, and observation of quantum path signatures in harmonic spectra from metal plasmas. Finally, we discuss future perspectives for HHG in laser plasma.

7.1. New Trends, Schemes, and Approaches in Plasma HHG

Plasma HHG has considerably matured during the last few years and continues to attract attention in various laboratories worldwide. Below, the most recent developments and some fresh approaches, experimental schemes, and ideas are described, which could push this field toward a dramatic improvement of the output characteristics of harmonics.

Intense HHG from plasma created from different carbon targets was demonstrated recently in [1]. High-order harmonic energy in the multimicrojoule range for each harmonic order from the 11th to the 17th harmonic was reported. By analyzing the target morphology and the plasma composition, the authors of [1] concluded that the intense harmonics from bulk carbon

185

targets originate from nanoparticles produced during ablation of the carbon-containing target. It was shown above that nanoparticles and films of C_{60} can generate harmonics that are more intense than those obtained from solid targets [2]. The disadvantage of using nanoparticle and film targets is the instability of the harmonic signal, which considerably varies shot-to-shot, and even disappears after a few laser shots if the target is not moved. In contrast, in [1] it was found for the first time that carbon bulk targets can generate intense harmonics, with intensity comparable to the nanoparticle or C_{60} film targets.

The SEM image of plasma debris from a carbon target revealed that the plasma plumes contain nanoparticles with sizes varying between 100 nm and 300 nm. It was therefore suspected that, during the interaction of the heating pulse with the carbon target, nanoparticles are formed in the plasma and are then pumped by the fundamental pulse to induce the generation of harmonics. The harmonic intensity using a bulk carbon target remained stable for several minutes, even without moving the target position. In creating a plasma for 5 minutes at the same place on the solid carbon target, the harmonic intensity did not decrease by more than 10%, while the plasma from nanoparticles decreased more than 90% after a few seconds.

It was also noted that the highest harmonic order obtained in the carbon target, unlike most other solid targets, does not exceed 21. According to the cutoff law defined by the three-step model, it was suggested that these harmonics are generated by neutral atoms, rather than ions as in the case of other solid targets.

Further developments of both stability of carbon-containing plasma harmonics and their enhanced yield were reported in [3]. The importance of these parameters is defined by the applicability of converted radiation for various needs. Many efforts have been dedicated to improvements of these characteristics during the long history of harmonic generation in gases. Multimicrojoule harmonics have been generated by energy scaling of gas HHG under highly optimized conditions [4, 5], which however has basically pushed gas HHG to its limits. Therefore, there is an urgent need to search for methods to generate even more intense harmonics and attosecond pulses. For these purposes, gas clusters [6] and plasma produced from nanoparticle targets [2, 7] can be used to increase the intensity of harmonics. In the former case, microjoule intense harmonics have been demonstrated at wavelengths of 50–90 nm.

However, as was mentioned above, nanoparticle targets have the problem of a rapid decrease in HHG intensity with consecutive shots, which prevents us from using them in applications. These exotic targets are also not always available in abundance. A new approach has been reported recently, which showed that highly efficient and relatively stable high-order harmonics could be generated from a target that is readily available in the home: the pencil lead of a mechanical pencil [3]. Measurements of the harmonic energy generated from plasma produced from pencil lead and comparison with the harmonics produced from C_{60} nanoparticles, which have proved to be one of most efficient media for plasma HHG, showed the advantages of harmonics from the former medium. The important advantage of using a pencil lead target is the shot-to-shot stability of the harmonic intensity over a sufficiently large number of shots.

To understand the uniqueness of the pencil lead plasma, the authors of [3] used a scanning electron microscope to analyze the ablated material debris deposited on silicon substrates that were placed close to the ablation plume. The SEM image (Fig. 7.1) revealed that the plasma created from the pencil lead target contains nanoparticles whose mean size was in the range of 200 nm. They therefore suspected that the ablation of the pencil lead target by the heating

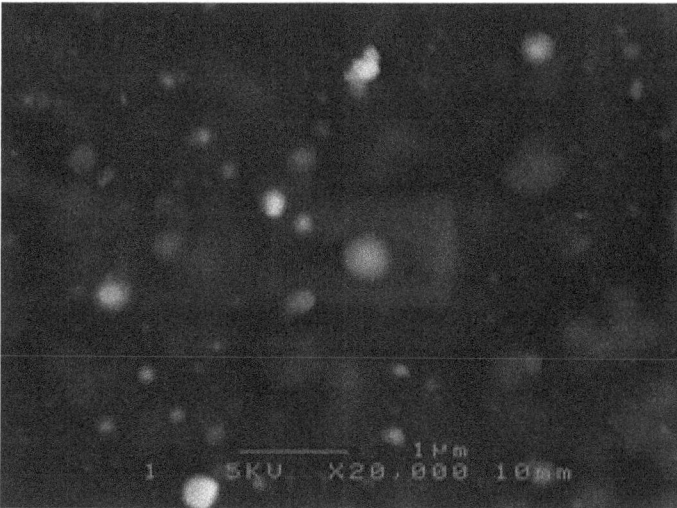

Fig. 7.1   SEM image of pencil lead plasma deposition. Reprinted from [3] with permission from American Institute of Physics.

pulse causes nanoparticles to be formed in the plasma, which in turn leads to generation of the intense harmonics.

From the experimental observations of stronger harmonics than in the case of fullerene plasma and the morphology of plasma debris, the authors of [1, 3] inferred that the origin of extremely strong harmonics from pencil lead and carbon plasmas is similar to that described for nanoparticle targets. The presence of nanoparticles in the plasma deposition suggests that neutral atoms of nanoparticles are the main source of intense harmonics from the pencil plasma. An explanation for intense harmonic generation from nanoparticles could be the higher concentration of neutral atoms due to the presence of nanoparticles. Unlike single atoms and ions, whose density quickly decreases due to plasma expansion, nanoparticles retain densities that are close to the material's solid state. Combined with the higher harmonic efficiency of neutral atoms compared with their ions, the neutral atoms within the nanoparticle could generate high-order harmonics efficiently. A conversion efficiency of $\sim 10^{-4}$ was estimated for the harmonics in the plateau range.

The important issue of HHG from plasmas is related to the characteristics of the generated harmonics. Whilst the conversion efficiency issue has been taken seriously during recent developments of this technique, which led to a considerable enhancement of harmonic pulse energy, no temporal characterization of plasma harmonic pulses has been performed up to recent times. This is a crucial element for the applications of this new source of XUV radiation. It should not be taken for granted that this harmonic emission has a good attosecond structure. Indeed, generation in plasma induces many sources of distortion: the higher electron densities and gradients will affect generation through phase matching and may result in distortion of both the harmonic spatial phase-front and the spectral phase. Furthermore, the temporal characterization itself raises problems, such as probe beam distortions, target deterioration, and instability of harmonic intensity.

The first measurements of attosecond emission generated from underdense plasma produced on a solid target were reported in [8]. The authors generated high-order harmonics of a femtosecond Ti:sapphire laser focused in a weakly ionized underdense chromium plasma. Characterization of the plasma attosecond emission was performed using the RABITT technique [9]. Measurement of the harmonic spectral amplitude and phase allowed direct access to the attosecond structure through a Fourier transform. The amplitude of each harmonic was easily given by the amplitude of the main photoelectron lines corrected for the ionization cross section. The relative phase between neighboring harmonic orders was accessed through two-photon $XUV + IR$ ionization of the target gas. When the dressing IR beam is superimposed on the XUV beam in the argon gas, sidebands appear in the photoelectron spectrum between the main lines. They correspond to two-photon transitions: absorption of a harmonic photon $q\omega_0$ accompanied by either absorption or stimulated emission of a laser photon ω_0. Since two coherent quantum paths lead to the same sideband, interferences occur that result in an oscillation of the sideband amplitude as the delay τ between the IR and harmonic field is scanned with sub-IR-laser-cycle resolution. The phase of the oscillation is the phase difference between the two interfering channels. The phase difference $\varphi_q - \varphi_{q+2}$ between two consecutive harmonics can then be extracted, readily giving the group delay, also called the emission time, $t_e = \partial\varphi/\partial\omega_{q+1} \approx -(\varphi_q - \varphi_{q+2})/2\omega_0$.

From the phases φ_q, obtained by integrating the emission times, and the amplitudes A_q of the harmonic orders, one can reconstruct the temporal intensity profile $I(t)$ according to:

$$I(t) = \left| \sum A_q \exp[-iq\omega_0 t + i\varphi_q] \right|^2 \qquad (7.1)$$

The result for the measured five harmonic orders $q = 11$ to 19 is shown in Fig. 7.2. The reconstructed temporal profile of the harmonic emission forms an attosecond pulse train, with each pulse of $300\,as$ duration (full width at half maximum). Assuming all five harmonics to be in phase, one can obtain the shortest pulses possible, i.e. the Fourier-transform limited pulses. The corresponding duration is $\tau = 285\,as$. The measured duration of $300\,as$ is thus only 1.05 times the Fourier-transform limited duration.

These studies allowed measurement of the spectral phase of the harmonic emission from chromium plasma. It exhibited an unexpectedly low group delay dispersion, which can be explained by the fact that the measured harmonic orders belong to the cutoff region of the Cr^+ HHG spectrum. Considering the potentially higher conversion efficiency when using higher generating intensity, these results indicate that plasma harmonics could become a source for generating intense attosecond pulses. Indeed, these measurements show that generation and propagation in the plasma do not significantly affect the group delay dispersion resulting from the single-ion response. The results

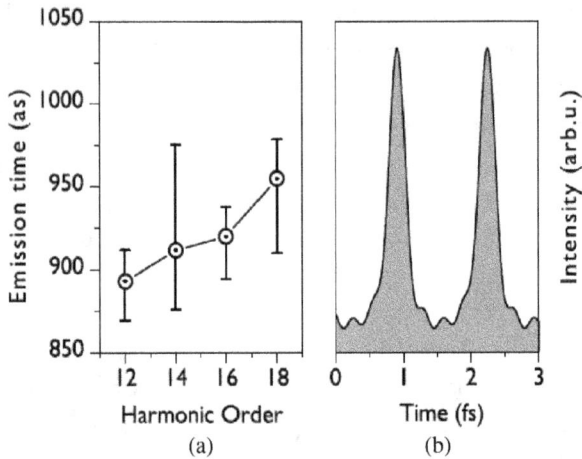

Fig. 7.2 (a) Emission times and (b) temporal profile corresponding to the 11th to 19th harmonic orders. Reprinted from [8] with permission from Optical Society of America.

predicted that generation at the saturation intensity of the Cr^{2+} ions, i.e., at 4×10^{14} W cm^{-2}, should both extend the plateau and result in a plateau group delay dispersion of 5300 as^2, low enough to generate intense 110 as pulses if 12 harmonic orders are selected. One can note that recent studies of plasma harmonic pulse duration showed the closeness of the temporal parameters of the driving pulse and converted radiation [10]. Furthermore, the application of a double optical gating allowed the generation of continuum harmonics in the case of carbon plasma, which indirectly points to the ultrashort nature of converted XUV pulses [11].

In the standard scheme of gas HHG, an ultrafast laser pump beam at intensities around 10^{14} W cm^{-2} is focused into a gas jet, generating high harmonics. The yield of such schemes is inevitably limited by dispersion in the medium. Across a distance equal to the coherence length, a phase mismatch of π grows and causes destructive interference between the pump and high harmonic beams. This process is a major limitation on the conversion efficiency of HHG. Quasi-phase matching (QPM) is a well-known approach for the phase mismatch problem [12]. In QPM, the medium is modulated with a coherence length period so that pump phase or harmonic emission is changed to prevent the destructive interference caused by the phase mismatch.

For HHG in the XUV regime and beyond, dispersion in the gaseous medium can be mostly attributed to free electrons generated by laser ionization of the medium. Under this assumption, the coherence length (at $0.8\,\mu m$ wavelength) is given (in meters) by [13] $L_c \propto 10^{15}/qN_e$ where N_e is the free-electron density (per cubic centimeter) and q is the harmonic number. Recently QPM was realized by using multiple gas jets whose pressure and separation were properly controlled [13]. However, the realization of this technique is limited by geometrical constraints on the number and minimal separation of the jets, as well as waveguide construction.

It has been proposed that the same procedure could be carried out for plasma HHG [14] using a simple method for fabricating numerous plasma jets tailored for HHG, relieving technical restraints on the dimensions of the jets and their periodicity. In this scheme the jets are produced by ablation of a microlithographic periodic stripe pattern (Fig. 7.3). Cylindrical plasma jets formed by ablation extend the lithographic pattern into the space above the target, creating a row of narrow plasma jets of different material composition. The efficiency of HHG in plasma has been demonstrated to vary considerably

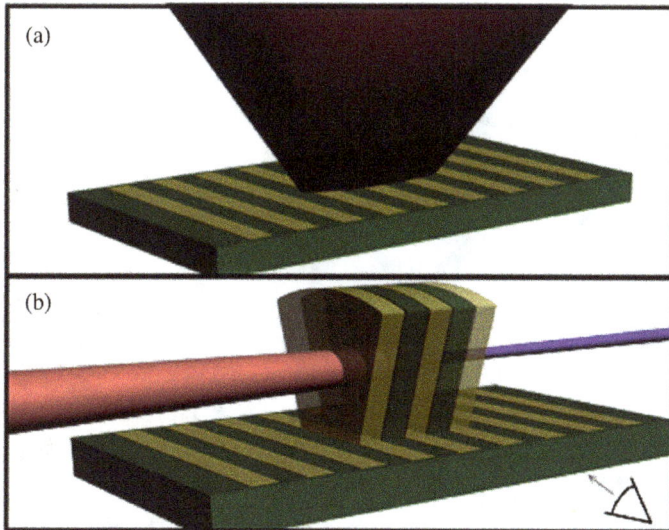

Fig. 7.3 System schematics: (a) the lithographic pattern hit by a relatively low-intensity laser beam, and (b) the formed plasma jets in which the high-intensity laser pump facilitates HHG. Reprinted from [14] with permission from American Institute of Physics.

with atomic composition [15], and the periodic change in this efficiency enables QPM-HHG.

The electron density and plasma structure that the HHG pump laser is incident upon can be adjusted (much as in using standard gas jets) by a multitude of control parameters. These parameters are: structure and material composition of the ablated target, the delay between the ablation and the pump incidence, the distance and angle at which the HHG pump is incident, and the ablating laser's parameters.

The results of [14] demonstrate a simple method for generation of periodic plasma structures by ablating a lithographic pattern. By passing a high-intensity laser pulse through such plasma patterns, suitable conditions for the QPM required for HHG can be created. These measurements suggest that such conditions exist around 140 to 180 ns after initiation of plasma by the ablating laser pulse (Fig. 7.4). Within this temporal window the plasma jets are several hundred micrometers wide and have relatively uniform temperature and a

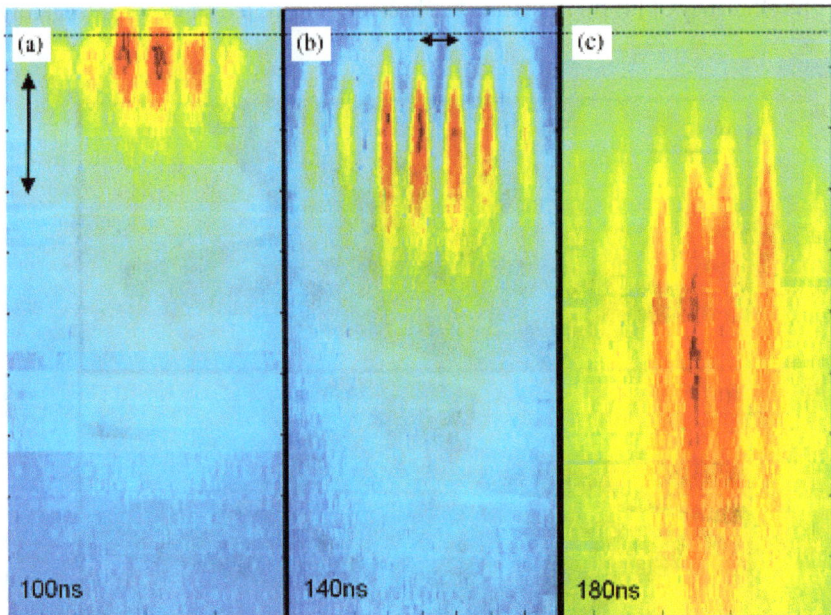

Fig. 7.4 Monochromatic imaging of the plasma jets at different times ((a) 100, (b) 140, and (c) 180 ns) using a 30 ns gate. The dashed line marks the target's surface, the double arrows are 200 μm scale. Reprinted from [14] with permission from American Institute of Physics.

relatively low electron density of $\sim 10^{17}$ cm^{-3}, whereas at later times the plasma structure begins to fade. Examining Fig. 7.4 in depth shows that the modulation of the plasma density is significant, while the authors suggest that much finer periodicities suitable for generation of higher harmonics could be obtained by using finer lithography in preparation of the target. They have demonstrated the feasibility of a robust scheme for tailoring plasma structures with control over material composition, temperature, and density (of both free electrons and neutrals), through ablation of specifically prepared lithographic targets, which can support QPM-HHG.

7.2. High-Order Harmonic Generation in Carbon Nanotube-Containing Plasma Plumes

Carbon nanotubes have remarkable electronic and optical properties due to their particular structure combining one-dimensional solid-state charac-teristics with molecular dimensions. While their structure has extensively been studied by means of transmission electron and scanning tunneling microscopy, only a few experimental studies have been reported on their nonlinear optical properties. Single-walled CNTs have a synthetic structure, and from a structural point of view they are graphene sheets rolled up into rather long cylindrical tubes [16]. The resulting low dimensionality determines most of the electronic properties relevant for the nonlinear optical response of single-walled CNTs. The quantum confinement of the delocalized p electrons in single-walled CNTs along the axis of symmetry of the cylinder, with the consequent minimization of the p–s electron mixing effects, is expected to enhance the third-order nonlinear optical response of CNTs with respect to other carbon-based structures such as fullerenes [17].

Nonlinear optical studies of CNTs using conventional methods have been limited for a long time to second- and third-harmonic generation of laser radiation [18–21], and measurements of third-order susceptibility [22]. The reported measurements were performed using both commercially available carbon nanotubes and samples of CNTs grown by a catalyst-free method. Third-harmonic generation was observed in both samples, while second-harmonic generation was observed only on the sample grown by the catalyst-free method. Evidence for second-order nonlinear optical activity in CNTs has also been demonstrated in a second-harmonic generation experiment performed at femtosecond timescale [23]. These works revealed the low-order nonlinear optical properties of nanotubes. However, no harmonic generation above third order in CNT media was reported until the ablation methods for HHG were introduced. It is of interest to study the higher-order harmonics in CNTs under monochromatic intense laser fields. The plasma plume method is perhaps the only way by which one can study HHG from CNTs. The main focus of the studies discussed below was the interaction of intense laser fields with the nanotubes in specially prepared laser plasma, which led to efficient harmonic generation in the XUV range using the two-color pump [24].

It was discussed earlier that by adding a second field to the fundamental one the harmonic spectrum exhibits many new features. For example, with the combination of a fundamental field and its second-order harmonic, both even and odd harmonics can be generated by sum and difference frequency mixing [25–32]. Some harmonics that cannot be observed in the monochromatic case can be produced in a bichromatic field. The enhancement of HHG has been experimentally observed both for parallel and perpendicular linearly polarized two-color fields. Two orders of magnitude intensity enhancement has been predicted theoretically in this HHG configuration, which has yet to be confirmed experimentally in gas HHG studies. In the meantime, as highlighted in previous chapters, it has been demonstrated for atom/ion-containing plasmas that intense odd and even harmonics can be generated from various ablated targets [33, 34]. So it would be interesting to combine the two approaches (i.e., two-color pump and CNT-containing plasma) to analyze the harmonics from nanotubes at favorable conditions of HHG.

The experiments with CNTs were performed using a 50 fs pulse and its second harmonic. The second-harmonic conversion efficiency was maintained at 8%. The polarizations of second-harmonic and fundamental fields were orthogonal due to type I phase matching in BBO. A laser ablation technique was used to produce a plasma plume containing CNTs. Most experiments were carried out using CNT powder glued onto glass substrates. In the case of CNT targets, an extended cutoff with harmonics up to the 29th order was observed. This was the first experimental observation of HHG in CNTs. The CNTs showed that they are stable against fragmentation in intense laser fields, which is probably due to the very fast distribution of excitation energy among the large number of carbon atoms present in a CNT, analogous to the fullerenes.

To understand the origin of the harmonic emission in CNTs, its dependence on the polarization of the driving pulse was analyzed. This also enables one to differentiate between plasma emission and the HHG process. As mentioned earlier, HHG is highly sensitive to laser polarization (linear or circular), since the trajectories of the recolliding electrons are altered significantly by the polarization, thereby inhibiting the recombination process, leading to decrease in HHG. The harmonic signal abruptly disappeared with change of polarization from linear to circular. Figure 7.5a shows the harmonic spectra obtained from a CNT-containing plasma plume in the cases of linear and circular polarization of the fundamental radiation using the single-color pump scheme.

Insertion of a 1 mm thick second-harmonic crystal in the beam path after the focusing lens led to generation of enhanced high-order harmonic yield and the appearance of even and odd harmonics with approximately equal intensities in the case of the lowest harmonic orders. An enhancement of HHG efficiency was observed in the two-color case compared to the single-color 800 nm pump (Fig. 7.5b). The even harmonics of the fundamental radiation up to the 16th order were obtained in these studies. The variation of laser chirp led to a change of harmonic generation efficiency of the odd and even harmonics. This was accompanied by tuning of the harmonic wavelengths, as well as an increase in the harmonic efficiency for negatively chirped laser pulses. The estimated efficiency of the harmonics in the range of 30–70 nm was 8×10^{-6}. It was found experimentally that the intensity of the harmonics from CNT-rich plumes was stronger compared to those generated from the plasma rich in single particles, created on the surface of bulk metal targets under the same experimental conditions.

The morphological characteristics of CNT targets prior to laser ablation were analyzed and compared with the ablated material debris deposited on a copper grid and carbon film. The diameter of the CNTs glued on the substrates was 3–6 nm, with length varying from 0.3 to 10 μm. The debris from the plasma plumes was studied at various pump pulse intensities. At a pump pulse intensity in the range of 3×10^9–2×10^{10} W cm^{-2}, CNTs were observed to be deposited intact. Figure 7.6 shows a TEM image of the deposited CNTs ($I_{pp} = 7 \times 10^9$ W cm^{-2}). No harmonics were observed during ablation of

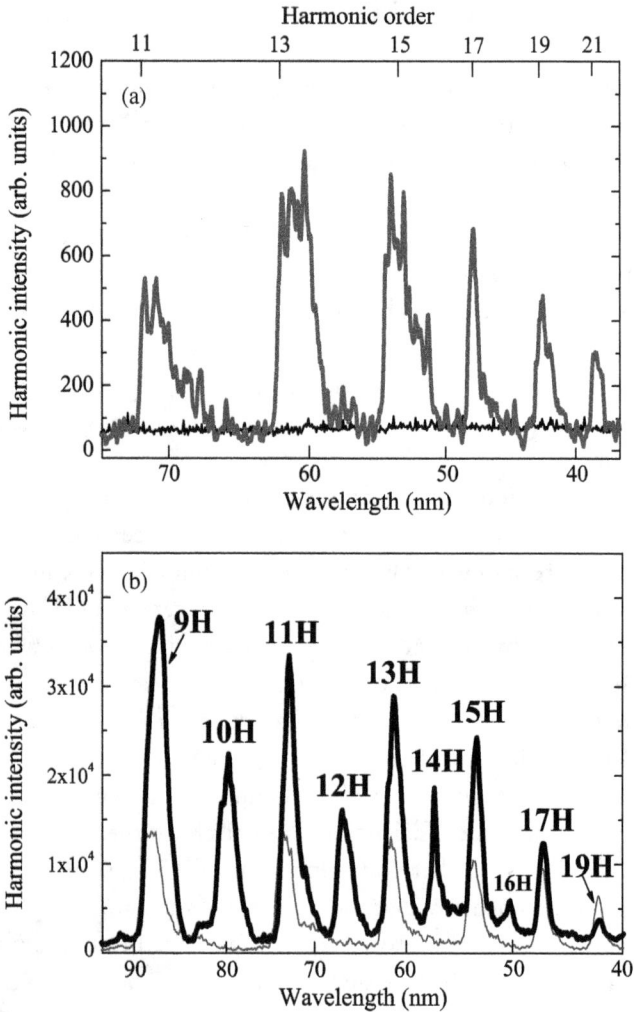

Fig. 7.5 (a) Harmonic generation in CNT plasma in the case of linear polarization (thick curve) and circular polarization (thin curve) of the probe radiation. (b) Comparison of harmonic generation from a carbon nanotube plasma plume in the cases of the single-color pump (thin curve) and two-color pump (thick curve) schemes. Adapted from [24] with permission from American Physical Society.

pure PMMA, superglue, or the glass substrate alone, without CNTs. The above-described morphology and HHG results indicate the ability of CNTs to survive strong excitation. From the above observations, one can conclude that CNTs are responsible for HHG.

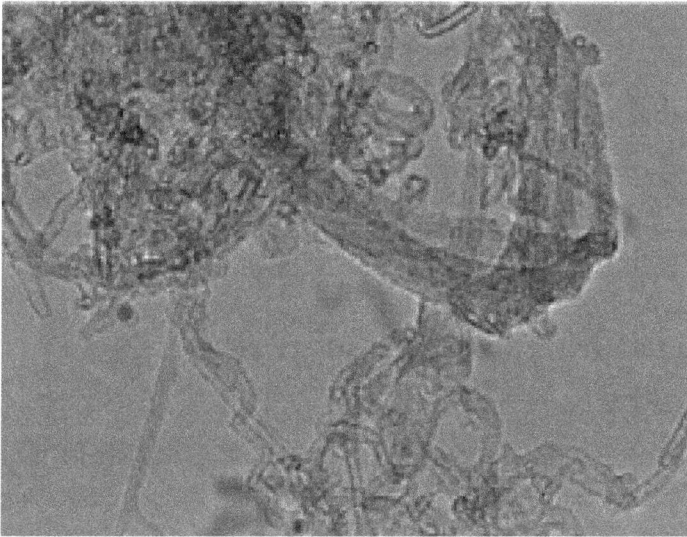

Fig. 7.6 TEM image of the deposited CNTs.

7.3. Destructive Interference of Laser Harmonics in Mixtures of Various Emitters in Plasma Plumes

Interference plays a noticeable role in various physical processes. It can cause both enhancement and decrease of the response of a medium to the influence of various forces. These features appear mostly during optical processes, in particular nonlinear optical phenomena in the presence of strong light fields. The strongest influence of interference can be observed in the case of coherent phenomena, when the phase of one optical process interferes with the phase of another process. One of the interesting examples of this phenomenon is the coherent superposition of two or more processes of HHG of the laser radiation emitted by particles of different physical origin, which are mixed and probed by femtosecond laser pulses. The important factor of the harmonic yield in this case is the relative phase of the harmonics originating from two (or more) different emitters.

A few studies have been reported on the interference of the HHG emitted by successive sources or mixtures of different gases [35–38]. The first evidence of destructive and constructive interference in a mixed gas of helium and neon, which can facilitate the coherent control of HHG, was reported in [36]. It was shown that, while in most cases this phenomenon leads to a decrease of harmonic conversion efficiency due to destructive interference of two HHG processes emitted from the two gas samples of different origin, there exist some spectral ranges where HHG can be enhanced through the constructive interference of these processes. The observed interference modulation was attributed to the difference between the phases of the intrinsically chirped harmonic pulses from helium and neon. From this point of view, it would be interesting to analyze the interference of harmonic generation in plasma plumes containing particles of different origin.

In the case of plasma formation, one can easily achieve such a mixing of particles possessing different physical properties (atomic number and weight, first and second ionization potentials, etc.). In [39], harmonic generation in a mixture of silver and gold nanoparticles was analyzed and compared with HHG in both these clusters contained singly in plasma plumes. Below we briefly describe some findings of this research.

The preparation of nanoparticle-containing targets was accomplished by using commercial nanoparticle suspensions. The initial molar concentration of the silver and gold nanoparticles in the suspensions was 10 mM. The solvents for these suspensions were water and n-propanol. The nanoparticles were protected against aggregation by adding polyethyleneimine. The silver and gold nanoparticle suspensions were dried on glass substrates and placed inside a vacuum chamber to ignite the plume by laser ablation of these structures. These two suspensions were also mixed at different volume ratios and then dried onto glass substrates. The concentration of silver and gold nanoparticle mixtures in the dried suspension was varied in the ranges of 1:1 and 1:3. These targets were used for analysis of the interference processes caused by simultaneous HHG using the nanoparticle-emitting centers. The configuration of the experiment using mixed targets (silver and gold nanoparticles), as well as double targets (bulk silver and gold), is presented in Fig. 7.7.

The spectra of harmonics generated from the silver and gold nanoparticles are presented in Figs. 7.8a and 7.8b. Here also are shown the harmonic spectrum obtained from the mixture of nanoparticles at the volume ratio of 1:3 (Fig. 7.8c). Interference modulation is observed in the spectra from the mixed plasma. In that case the harmonics over the entire spectral range were suppressed. This decrease of HHG efficiency was attributed to the

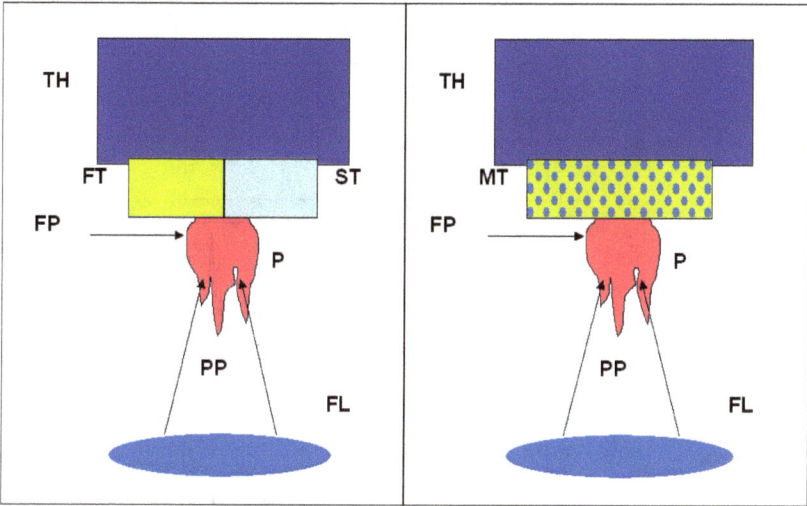

Fig. 7.7 Target configurations for the measurements of interference between the harmonics generated from different emitters. (a) Double-target scheme: TH, target holder; FP, femtosecond pulse; PP, picosecond pulse; P, plasma plume; FL, focusing lens; FT, first target; ST, second target. (b) Mixed-target scheme: as for (a) with MT, mixed target. Adapted from [39] with permission from Springer Science+Business Media.

destructive interference between the harmonics generated from the silver and gold nanoparticles.

The term "modulation" itself points to the conditions when one parameter or process (in our case, harmonic generation from one group of emitters) modulates the overall response of the system. So, concerning those studies, this term refers to the conditions when the harmonic yield is modified (enhanced in the case of constructive interference and decreased in the case of destructive interference). In the presented case, only a decrease of harmonic yield over the entire spectral range was observed, which means that destructive interference is the main mechanism governing the entire harmonic output from the mixed plasma. Over-excitation of the target containing silver and gold nanoparticles by the heating pulse led to disruption of HHG. In that case, the spectrum of radiation contained only plasma emission from the nanoparticles and ions (Fig. 7.8d).

Note that when two separate bulk targets (silver or gold) were used, and the so-called double-target configuration was applied for experiments [40, 41], no considerable change in harmonic yield in the double case compared with the

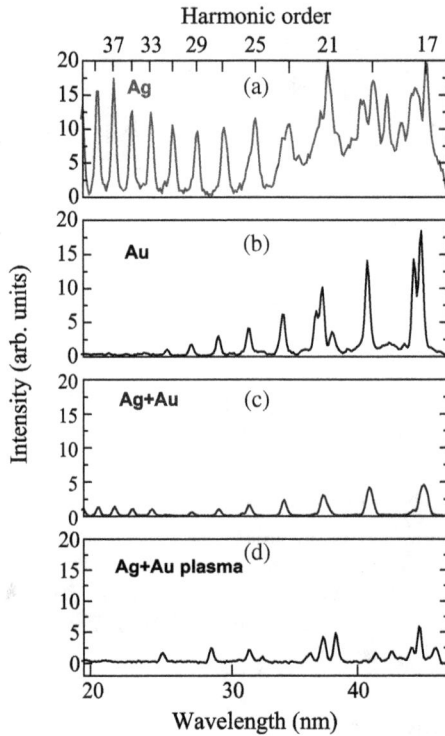

Fig. 7.8 Harmonic spectra obtained in the plasmas containing (a) silver, (b) gold, and (c) silver + gold nanoparticles. In the third case a mixed-target configuration (Fig. 7.7b) was used. (d) Plasma spectrum observed at over-excitation of the target containing the mixed dried nanoparticle suspension. Reprinted from [39] with permission from Springer Science+Business Media.

two single cases was observed. In the double case, the targets contained silver and gold plates connected to each other (Fig. 7.7a). The plasma was created simultaneously on the two targets. The size of the laser plume in the double-target configuration was the same as in the case of single targets (~1 mm). The harmonic spectrum in that case was approximately equal to the convolution of two harmonic spectra produced separately from silver and gold ions ablated on the surfaces of bulk targets. Figure 7.9 shows the harmonic spectra from the silver (curve 1), gold (curve 2), and silver + gold (curve 3) plasmas produced on the surfaces of bulk targets. In the double case, no interference effects can be expected since the distance between the emitters of different origin (a few hundreds of micrometers) was considerably longer than that in the case

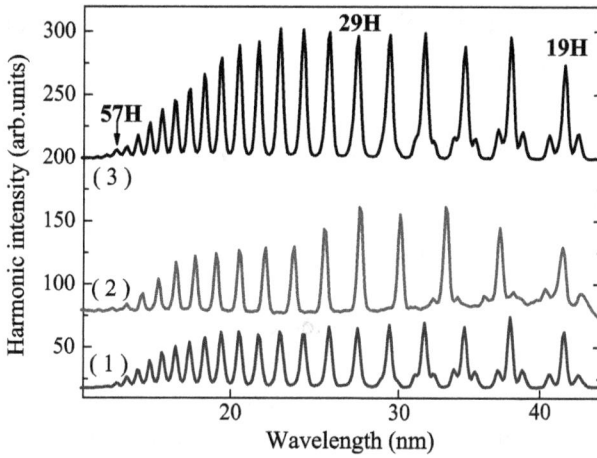

Fig. 7.9 Harmonic spectra from the (1) silver, (2) gold, and (3) silver + gold plasmas produced on the surfaces of bulk targets. In the third case, a double-target configuration (Fig. 7.7a) was used. Reprinted from [39] with permission from Springer Science+Business Media.

of mixed emitters (of the order of hundreds of nanometers) in the plasma containing silver and gold nanoparticles.

In the case of a homogeneously mixed medium, one can expect an influence of the excursion time of ejected electrons before recombination with parent particles on the interference processes of plasma HHG. If we consider a bulk target initially containing different materials, the overall plasma harmonic efficiency should depend on the interference of the harmonics generated from the ingredients of the laser ablation plasma. In that case, the efficiency of harmonics is limited by dephasing of the atomic dipole oscillators driven at different positions in the generation medium. Since HHG is a coherent process, interference can considerably influence the harmonic efficiency. Considering the contributions in the plume from two different induced dipole moments, the interference should be related to their different dipole phases. In this case, the interference mechanism would occur even on a microscopic scale, since the medium is isotropic and uniform. It is inferred from those results that dipole phase interference is the proper explanation of the observed phenomenon. As an interference mechanism, it is expected that some harmonics suffer destructive superposition, but other harmonics may show constructive interference, as was found in [36] for a gas mixture. However, in the plasma HHG case, almost entire homogeneous degradation of the whole HHG spectra was observed, with no sign of constructive interference in the analyzed spectral range (25–50 nm).

7.4. Generation of Broadband Harmonics from Laser Plasma

Various methods allow considerable modification of the properties of harmonic radiation. Among them, phase modulation of laser radiation can easily tune the spectrum and intensity of harmonics. Self-phase modulation is widely used to generate additional frequencies in laser pulses. In particular, insertion of an optical medium (such as a glass slab) in the path of a laser pulse leads to moderate spectral broadening of the laser pulse by SPM. Propagation of a focused intense laser pulse through air can also cause significant SPM-induced broadening of a driving pulse and remarkably change the harmonic pattern during HHG in laser plasma using this radiation. In this section, harmonic spectra obtained during HHG in silver plasma using narrowband laser pulses (~10 nm) are analyzed. Very broad harmonics were observed in the case of focusing of the laser radiation in air prior to the generation of harmonics in the laser plume [42].

The schemes for laser ablation were analogous to those used in plasma HHG. The difference was mostly related to the characteristics of the femtosecond pulse. In the first case (see setup in Fig. 7.10a), the conventional scheme, where a compressed femtosecond pulse propagates through the plasma, was used. This driving pulse (central wavelength 790 nm and bandwidth 10 nm) had 12 mJ energy and pulse duration of 120 fs after propagation through a delay line and compressor stage. A 60 ns delay between the pump and driving pulses was used in these studies. The driving pulse intensity at the plasma plume was varied between 5×10^{14} and 2×10^{15} W cm^{-2}. Figure 7.10a shows the rough image of harmonic spectra obtained using a CCD camera after propagation of the focused driving pulse through the "optimal" silver plasma. This term refers to the plasma conditions when the highest intensity of harmonics was achieved. The harmonics up to 51st order are clearly seen in these spectral images. Note that the conversion efficiency of shorter-wavelength harmonics was higher compared with low-order harmonics, which was caused by better phase matching conditions in the former case (see also [15, 43]). The scheme of strong SPM-induced HHG experiments is presented

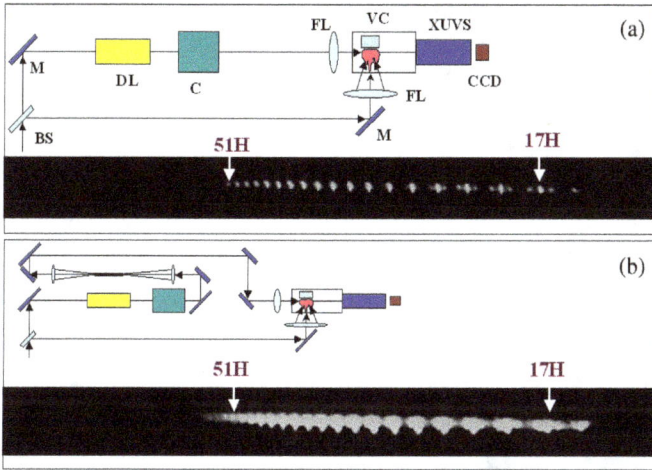

Fig. 7.10 (a) Rough pattern of harmonic spectra from silver plasma in the case of modulation-free laser pulses, and experimental setup of harmonic generation in laser plasma: BS, beam splitter; M, mirrors; DL, delay line; C, compressor; FL, focusing lenses; VC, vacuum chamber; XUVS, extreme ultraviolet spectrometer; CCD, charge coupled device. (b) Rough pattern of harmonic spectra obtained in the case of harmonic generation using modulated laser pulses propagating through the telescope and filaments, and experimental setup of these experiments. Reprinted from [42] with permission from American Institute of Physics.

in Fig. 7.10b. After passing the compressor stage, the laser beam was propagated in air through the 1:1 telescope consisting of two 2000 mm focal length spherical lenses. The value of the Rayleigh length before filamentation for the lenses used was 47 mm. In the focal area, 300 mm long filaments and white light generation were observed. The laser radiation then returned to the vacuum chamber and was focused in the silver plasma. At these conditions, generation of extremely broadened harmonics was obtained (Fig. 7.10b). One can see here stronger harmonics compared with the case presented in Fig. 7.10a. These HHG experiments were carried out at conditions analogous to the previous ones, i.e., the same energy of the driving pulse before the vacuum chamber was maintained for these comparative studies, together with analogous plasma conditions.

The spatial, spectral, and temporal characteristics of the 790 nm radiation propagating through the telescope and creating air filaments were analyzed and compared with modulation-free laser radiation. Figure 7.11 shows the spatial shapes of the beams in front of the vacuum chamber in these two cases. The

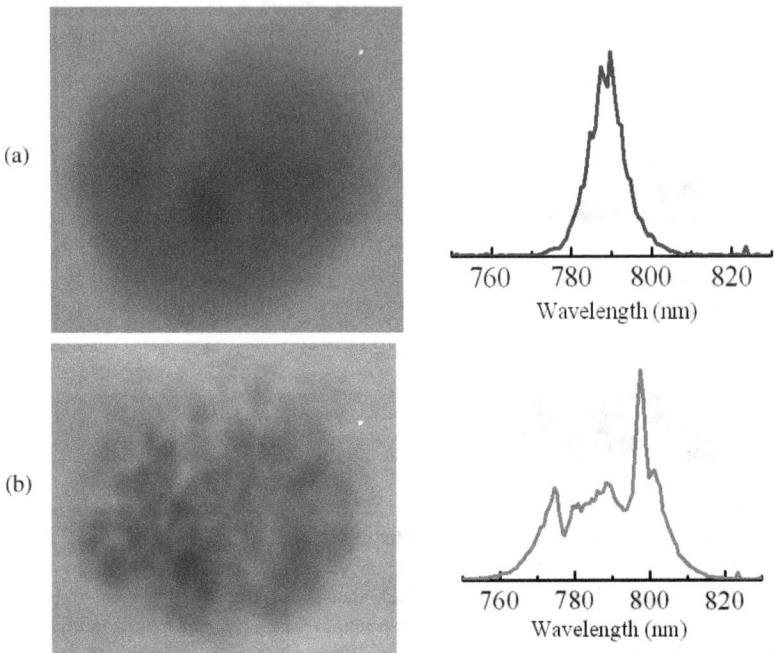

Fig. 7.11 Spatial shapes of (a) laser beam after the compressor and (b) laser beam after propagation through the telescope and filamentation. The corresponding spectral distributions of the laser pulses for these two cases are also presented. Reprinted from [42] with permission from American Institute of Physics.

initial shape of the compressed driving beam was relatively smooth, without considerable variations of laser intensity along the beam (Fig. 7.11a). The spatial shape of the laser beam creating filaments and propagating through air, even after some smoothening during the long pass, showed hot spots created due to the filamentation of different parts of the laser beam (Fig. 7.11b). The pulse durations in front of the vacuum chamber for these two cases were 120 and 195 fs, respectively.

Strong SPM and white light generation in the area of filament formation also caused a considerable change of the spectral shape of the driving laser pulse. The spectra of the driving pulse after the compressor and after propagation of air filaments differ considerably from each other. While the spectral width of the laser pulse after the compressor stage always remains in the region of 10 nm, a dramatic increase of spectral width of the laser radiation propagating through the telescope was attributed to the above-mentioned nonlinear optical processes. The additional frequencies generated along the z-axis in that case are given by

$$\omega = \omega_0 - k_0 z n_2 \mathrm{d}I/\mathrm{d}t \qquad (7.2)$$

where n_2 is the nonlinear refractive index of the medium, k_0 is the wave vector of the laser pulse, ω_0 and ω are the initial and resultant frequencies of the laser pulse, and I is the intensity of the laser pulse. The overall spectral width of the laser pulse was increased from 10 to 35 nm (Fig. 7.11). Since the nonlinear refractive index of air $n_2 > 0$, the initial portion of the driving pulse ($\mathrm{d}I/\mathrm{d}t > 0$) generates red frequencies and the trailing portion of the laser pulse generates blue frequencies. One can note that the spectral components related to white light generation considerably diminished during propagation over the long pass from the second lens of the telescope to the vacuum chamber (~ 10 m) where the final spectral measurements of the laser pulse were carried out. This was due to the higher divergence of the white light components, which were partially removed from the overall spectra after propagation over a long distance, as well as the reflective properties of the mirrors used for directing the SPM-affected beam back to the vacuum chamber.

There have been a few studies of the variation of harmonic bandwidth due to spectral broadening of the driving radiation [44–46]. However, all these studies were addressed to the conditions when laser broadening is induced by the relatively "mild" influence of SPM on the harmonic properties. This term refers to the conditions when both spectral and spatial parameters of laser radiation moderately change during propagation through the nonlinear medium. In the reviewed work, the first observation and analysis of harmonic characteristics at the conditions when laser radiation significantly changes due to propagation in air, strong filamentation, and remarkable growth of spectral and spatial frequency components prior to the interaction with nonlinear medium was analyzed. The origin of the enhancement of harmonic efficiency at these conditions is not clear and requires additional study.

7.5. Quantum Path Signatures in Harmonic Spectra from Metal Plasma

High-order harmonic generation of laser radiation is a useful tool for analysis of various processes occurring in the time interval between ionization and recombination [47, 48]. In particular, electron continuum wavepackets following trajectories of different duration leading to the same emitted photon energy will acquire a different phase [49]. The resulting phase difference between the various contributions has been shown to reveal itself through interference effects between short and long trajectories [50].

Some peculiarities of quantum path interference in the case of ionized media were discussed in [51]. It was predicted that the effect of ionization can result in an asymmetric spectral shape of the interference pattern for different harmonic orders. Those simulations showed that the asymmetry of the harmonic spectrum in the case of ionized media is a result of ground-state depletion, which is a single-atom effect. The harmonic signal on the red side of the spectrum is reduced because the trailing edge of the generating laser pulse "sees" a medium that is already depleted by the leading edge.

One can note that until recent studies there had been no explicit confirmation of quantum path interference in the case of the harmonics generated in plasmas. The limitation of this observation was related to the spatial and temporal averaging under the usual macroscopic generation conditions, as well as insufficient laser stability, which smeared out the interference pattern during studies using the low (10 Hz) pulse repetition rate laser sources commonly used for plasma HHG. The availability of higher pulse repetition rate Ti:sapphire lasers has had a large impact on strong field research over the past decade. Below, we discuss studies of the quantum path signatures (QPS) of the two quantum paths contributing to harmonic emission in laser plasma plumes containing a sufficient amount of ionized medium [52]. The high pulse repetition rate (1 kHz) of the Ti:sapphire lasers used [53] allowed observation of the contributions of short and long electron trajectories in the case of plasma HHG.

Part of the uncompressed radiation from a Ti:sapphire laser (780 nm, 1.5 mJ, 20 ps, 1 kHz) was split from the beam line prior to the laser compressor stage and focused into the vacuum chamber to create plasma on the target surfaces (aluminum, copper, silver, manganese). The plasma plumes were moved over a range of positions with respect to the focal point of the probe pulse, to analyze the influence of the focusing geometry on harmonic divergence and trajectory isolation. The QPS was studied for different conditions of laser plasma and laser parameters and the role of ionization effects on the observed harmonic pattern was analyzed. Most of the experiments were carried out using aluminum plasma at a pulse delay between pump and probe of 26 ns. The QPS was observed in copper, manganese, and silver plasmas as well. In the first stage, the harmonic generation from the aluminum plasma was optimized by adjusting the position of the focus with respect to the plasma plume, the distance between target surface and femtosecond beam propagation axis, and the intensity of the heating picosecond pulse at the ablated area.

When selecting the short trajectory (plasma plume after the laser focus), spectrally narrow harmonics were generated as a result of the small frequency chirp of the short trajectory. No QPS was observed in that case. In contrast, both trajectories became phase matched for a plasma plume placed before the laser focus. Harmonics up to the 39th order were routinely obtained from aluminum plasma. A 5 mm aperture was then introduced in the probe beam in front of the vacuum chamber, which considerably changed the shape of the harmonic spectra. The brightness of the on-axis harmonics became much stronger than in the case of the apertureless femtosecond beam. This feature has previously been observed in multiple gas and plasma HHG experiments. This improvement in on-axis HHG efficiency is attributed to better phase matching conditions associated with the increase of the confocal parameter. The most important change between these two regimes (i.e., with and without the 5 mm aperture) was the appearance of rings around the harmonics. On either side of the central harmonic frequency, a parabolic shape of the harmonic spectral distribution became visible, where the red component of these parabolas was considerably weaker than the blue one, which could be explained by the above-mentioned assumption reported in [51]. The acquisition of each harmonic spectrum in this case was performed over 2000 laser shots. High-energy stability of the laser radiation was required to ensure that QPS was not smeared out due to intensity fluctuations. The same can be said of the necessity for stability of plasma density.

The relative contributions from the short and long trajectories can be adjusted by changing the position of the plasma plume with respect to the focal point, the chirp of the probe pulse, the plasma concentration, and the intensity of the laser radiation. Hence, QPS contrast can be maximized and a clear variation of the QPS with these parameters can be achieved. To analyze this process, harmonic spectra were measured at different positions of the plasma with respect to the focus of the femtosecond beam (Fig. 7.12).

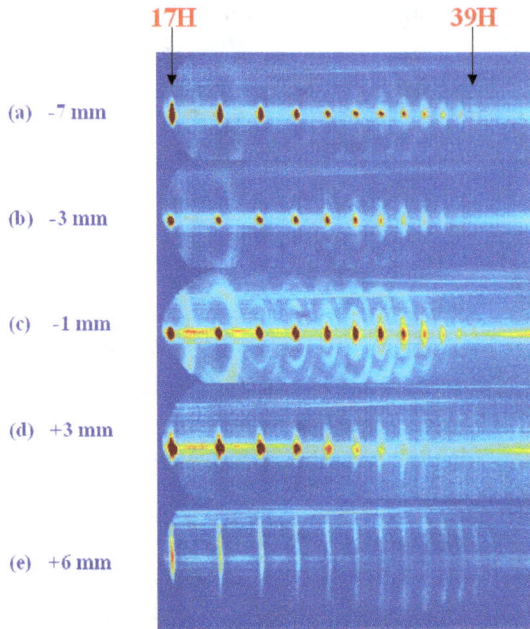

Fig. 7.12 Dependence of harmonic spectra on the position of aluminum plasma with respect to the focus of femtosecond radiation (negative values on the pictures correspond to focusing after the plasma plume). Harmonic orders are shown on the figure. Reprinted from [52] with permission from American Physical Society.

It has been shown that rings appear exclusively when the plasma is placed before the focus, in accordance with theoretical predictions [54] and experimental observations in gas HHG [50, 51]. At a sufficiently high intensity, rings from neighboring harmonics start to overlap. The maximum ring contrast was observed when the plasma was placed close to the focal point (Fig. 7.12c). In the case of HHG in noble gases, the pattern may be affected at the focus position if high ionization occurs (for low I_p level gases [51]). However, in the case of plasma plumes, where not only neutrals but also ions are involved in the HHG process, higher intensities can be considered and so observation of QPS patterns is also possible close to the focal position. For the highest harmonic orders, those in the cutoff region, modulations were not observed because only one (short) trajectory contributes to the emission. The rings around all other generating harmonics were obtained, while the longer-wavelength part of these rings was weaker than shorter-wavelength one. This asymmetry could be induced by further ionization of singly charged ions by the leading edge of the pulse, which occurs at $I_{fp} = (3\text{--}5) \times 10^{14}\,\mathrm{W\,cm^{-2}}$, followed by a change in phase matching conditions causing spectral distortions [51].

7.6. Summary and Future Perspectives of HHG in Laser Plasma

In this book, a review of the achievements of harmonic generation from laser plasma is presented. The most important studies are analyzed from the standpoint of their involvement in the development of effective coherent XUV sources. The earliest of those studies was presented in a review published in 2007 [55]. In recent years, the development of this technique has revealed many new approaches and achievements in harmonic generation from the plasmas, which were analyzed in the reviews [56, 57]. The most intriguing among them are harmonic generation using ablation of commercially available fullerenes and carbon nanotubes, application of high pulse repetition sources and ultrashort pulses for HHG in plasma plumes, observation of quantum path signatures in harmonic spectra from various plasmas, achievement of microjoule-level harmonic energies, and measurements of plasma harmonic pulse durations on the attosecond timescale.

Future developments in the application of this technique may include such areas as the seeding of plasma resonance harmonics in XUV free-electron lasers, plasma-induced harmonic generation using a few-cycle pulses, application of endohedral fullerenes for plasma HHG, comparative studies of gas- and plasma-induced harmonics, analysis of molecular structures through the study of harmonic spectra from oriented molecules in plasmas, the search and application of quasi-phase matching schemes in plasma plumes, the use of single harmonics for surface science, structural analysis of multi-dimensional formations in laser plasma, generation of strong combs and single attosecond pulses during plasma HHG, a quest for quasi-solid-state HHG, application of the double-target schemes for plasma formation, use of rotating targets for improvements of harmonic stability, application of infrared (1000–3000 nm) laser sources for extension of plasma harmonic cutoffs, analysis of plasma components through the HHG, etc. All the aforementioned achievements and perspectives unambiguously show the attractiveness of this method for coherent short-wavelength radiation generation and analysis of matter. These studies show that harmonic generation in plasma media has become a mature field of nonlinear optics, which allows both creation of efficient coherent XUV radiation sources and their applications.

The studies described in this book pave the way for development of a new method of materials studies using *laser ablation-induced high-order harmonic generation spectroscopy*, which can exploit the spectral and structural properties of various solid-state materials through their ablation and further propagation of short laser pulses through laser-produced plasmas and the generation of high-order harmonics. This approach will (i) open up new opportunities for high-order generation of ultrashort pulses in laser-produced plasma plumes, (ii) develop tools for studies of large molecules and clusters in the ablated conditions, and (iii) broaden the range of subjects for study compared with the presently used HHG in gases.

References

1. Elouga Bom, L.B., Petrot, Y., Bhardwaj, V.R. *et al.* (2011). Multi-μJ coherent extreme ultraviolet source generated from carbon using the plasma harmonic method, *Opt. Express*, 19, 3077–3085.
2. Ganeev, R.A., Elouga Bom, L.B., Abdul-Hadi, J. *et al.* (2009). High-order harmonic generation from fullerene using the plasma harmonic method, *Phys. Rev. Lett.*, 102, 013903.
3. Petrot, Y., Elouga Bom, L.B., Bhardwaj, V.R. *et al.* (2011). Pencil lead plasma for generating multimicrojoule high-order harmonics with a broad spectrum, *Appl. Phys. Lett.*, 98, 101104.
4. Hergott, J.-F., Kovacev, M., Merdji, H. *et al.* (2002). Extreme-ultraviolet high-order harmonic pulses in the microjoule range, *Phys. Rev. A*, 66, 021801.
5. Takahashi, E., Nabekawa, Y. and Midorikawa, K. (2002). Generation of 10-μJ coherent extreme-ultraviolet light by use of high-order harmonics, *Opt. Lett.*, 27, 1920–1922.
6. Donnelly, T.D., Ditmire, T., Neuman, T. *et al.* (1996). High-order harmonic generation in atom clusters, *Phys. Rev. Lett.*, 76, 2472–2475.
7. Elouga-Bom, L.B., Ganeev, R.A., Abdul-Hadi, J. *et al.* (2009). Intense multi-microjoule high-order harmonics generated from neutral atoms of In_2O_3 nanoparticles, *Appl. Phys. Lett.*, 94, 111108.
8. Elouga Bom, L.B., Haessler, S., Gobert, O. *et al.* (2011). Attosecond emission from chromium plasma, *Opt. Express*, 19, 3677–3685.
9. Paul, P.M., Toma, E.S., Breger, P. *et al.* (2001). Observation of a train of attosecond pulses from high harmonic generation, *Science*, 292, 1689–1692.
10. Haessler, S., Elouga Bom, L.B., Gobert, O. *et al.* (2012). Femtosecond envelope of the high-harmonic emission from ablation plasmas, *J. Phys. B: At. Mol. Opt. Phys.*, 45, 074012.

11. Pertot, Y., Chen, S., Khan, S.D. *et al.* (2012). Generation of continuum high-order harmonics from carbon plasma using double optical gating, *J. Phys. B: At. Mol. Opt. Phys.*, **45**, 074017.
12. Bahabad, A., Murnane, M.M. and Kapteyn, H.C. (2010). Quasi-phase-matching of momentum and energy in nonlinear optical processes, *Nature Phot.*, **4**, 570–575.
13. Hentschel, M., Kienberger, R., Spielmann, C. *et al.* (2001). Attosecond metrology, *Nature*, **414**, 509–514.
14. Sheinfux, A.H., Henis, Z., Levin, M. *et al.* (2011). Plasma structures for quasiphase matched high harmonic generation, *Appl. Phys. Lett.*, **98**, 141110.
15. Ganeev, R.A., Singhal, H., Naik, P.A. *et al.* (2007). Optimization of the high-order harmonics generated from silver plasma, *Appl. Phys. B*, **87**, 243–247.
16. Iijima, S. (1991). Helical microtubules of graphitic carbon, *Nature*, **354**, 56–58.
17. Flom, S.R., Pong, S., Bartoli, F.J. *et al.* (1992). Resonant nonlinear optical response of the fullerenes C_{60} and C_{70}, *Phys. Rev. B*, **46**, 15598–15601.
18. Liu, X. Si, J., Chang, B. *et al.* (1999). Third-order optical nonlinearity of the carbon nanotubes, *Appl. Phys. Lett.*, **74**, 164–166.
19. Wang, S., Huang, W., Yang, H. *et al.* (2000). Large and ultrafast third-order optical nonlinearity of single-wall carbon nanotubes at 820 nm, *Chem. Phys. Lett.*, **320**, 411–414.
20. Stanciu, C., Ehlich, R., Petrov, V. *et al.* (2002). Experimental and theoretical study of third-order harmonic generation in carbon nanotubes, *Appl. Phys. Lett.*, **81**, 4064–4066.
21. De Dominicis, L., Botti, S., Asilyan, L.S. *et al.* (2004). Second- and third-harmonic generation in single-walled carbon nanotubes at nanosecond time scale, *Appl. Phys. Lett.*, **85**, 1418–1420.
22. Lauret, J.-S., Voisin, C., Cassabois, G. *et al.* (2004). Third-order optical nonlinearities of carbon nanotubes in the femtosecond regime, *Appl. Phys. Lett.*, **85**, 3572–3574.
23. Konorov, S.O., Akimov, D.A., Ivanov, A.A. *et al.* (2003). Femtosecond optical harmonic generation as a non-linear spectroscopic probe for carbon nanotubes, *J. Raman Spectrosc.*, **34**, 1018–1024.
24. Ganeev, R.A., Naik, P.A., Singhal, H. *et al.* (2011). High order harmonic generation in carbon nanotube-containing plasma plumes, *Phys. Rev. A*, **83**, 013820.
25. Cormier, E. and Lewenstein, M. (2000). Optimizing the efficiency in high order harmonic generation optimization by two-colour fields, *Eur. Phys. J. D*, **12**, 227–233.
26. Kim, I.J., Kim, C.M., Kim, H.T. *et al.* (2005). Highly efficient high-harmonic generation in an orthogonally polarized two-colour laser field, *Phys. Rev. Lett.*, **94**, 243901.
27. Mauritsson, J., Johnsson, P., Gustafsson, E. *et al.* (2006). Attosecond pulse trains generated using two colour laser fields, *Phys. Rev. Lett.*, **97**, 013001.
28. Pfeifer, T., Gallmann, L. and Abel, M.J. (2006). Single attosecond pulse generation in the multicycle-driver regime by adding a weak second-harmonic field, *Opt. Lett.* **31**, 975–977.
29. Charalambidis, D., Tzallas, P., Benis, E.P. *et al.* (2008). Exploring intense attosecond pulses, *New J. Phys.*, **10**, 025018.
30. Kim, I.J., Lee, G.H., Park, S.B. *et al.* (2008). Generation of submicrojoule high harmonics using a long gas jet in a two-color laser field, *Appl. Phys. Lett.*, **92**, 021125.
31. Aubanel, E.E. and Bandrauk, A.D (1994). Orbital alignment and electron control in photodissociation products by two-colour laser interference, *Chem. Phys. Lett.*, **229**, 169–174.

32. Zuo, T., Bandrauk, A.D., Ivanov, M. *et al.* (1995). Control of high-order harmonic generation in strong laser fields, *Phys. Rev. A*, 51, 3991–3998.
33. Ganeev, R.A., Singhal, H., Naik, P.A. *et al.* (2009). Enhancement of high-order harmonic generation using two-color pump in plasma plumes, *Phys. Rev. A*, 80, 033845.
34. Ganeev, R.A., Singhal, H., Naik, P.A. *et al.* (2010). Systematic studies of two-color pump induced high order harmonic generation in plasma plumes, *Phys. Rev. A*, 82, 053831.
35. Seres, J., Yakovlev, V.S., Seres, E. *et al.* (2007). Coherent superposition of laser-driven soft-X-ray harmonics from successive sources, *Nature Phys.*, 3, 878–883.
36. Kanai, T., Takahashi, E.J., Nabekawa, Y. *et al.* (2007). Destructive interference during high harmonic generation in mixed gases, *Phys. Rev. Lett.*, 98, 153904.
37. Takahashi, E.J., Kanai, T., Ishikawa, K.L. *et al.* (2007). Dramatic enhancement of high-order harmonic generation, *Phys. Rev. Lett.*, 99, 053904.
38. Kanai, T., Takahashi, E.J., Nabekawa, Y. *et al.* (2008). Observing molecular structures by using high-order harmonic generation in mixed gases, *Phys. Rev. A*, 77, 041402.
39. Ganeev, R.A. and Kuroda, H. (2011). Destructive interference in the mixtures of different harmonic emitters inside the plasma plumes, *Appl. Phys. B*, 103, 151–159.
40. Suzuki, M., Ganeev, R.A., Ozaki, T. *et al.* (2007). Enhancement of two-color high harmonic by using two compound strong radiative transition ions in double-target scheme, *Appl. Phys. Lett.*, 90, 261104.
41. Suzuki, M., Ganeev, R.A., Baba, M. *et al.* (2008). Characteristics of high-order harmonic spectrum by using laser-ablated two targets combination, *Phys. Lett. A*, 372, 4480–4483.
42. Ganeev, R.A. and Kuroda, H. (2009). Extremely broadened high-order harmonics generated by the femtosecond pulses propagating through the filaments in air, *Appl. Phys. Lett.*, 95, 201117.
43. Ganeev, R.A., Baba, M., Suzuki, M. *et al.* (2005). High-order harmonic generation from silver plasma, *Phys. Lett. A*, 339, 103–109.
44. Tosa, V., Kim, H.T., Kim, I.J. *et al.* (2005). High-order harmonic generation by chirped and self-guided femtosecond laser pulses. II. Time-frequency analysis, *Phys. Rev. A*, 71, 063808.
45. Froud, C.A., Rogers, E.T.F., Hanna, D.C. *et al.* (2006). Soft-x-ray wavelength shift induced by ionization effects in a capillary, *Opt. Lett.*, 31, 374–376.
46. Ganeev, R.A, Singhal, H. Naik, P.A. *et al.* (2009). Variation of harmonic spectra in laser-produced plasmas at variable phase modulation of femtosecond laser pulses of different bandwidth, *J. Opt. Soc. Am. B*, 26, 2143–2151.
47. Lewenstein, M., Balcou, P., Ivanov, M.Y. *et al.* (1994). Theory of high-harmonic generation by low-frequency laser fields, *Phys. Rev. A*, 49, 2117–2132.
48. Marangos, J.P., Baker, S., Kajumba, N. *et al.* (2008). Dynamic imaging of molecules using high order harmonic generation, *Phys. Chem. Chem. Phys.*, 10, 35–48.
49. Toma, E.S., Antoine, P., de Bohan, A. *et al.* (1999). Resonance-enhanced high-harmonic generation, *J. Phys. B: At. Mol. Opt. Phys.*, 32, 5843–5852.
50. Zaïr, A., Holler, M., Guandalini, A. *et al.* (2008). Quantum path interferences in high-order harmonic generation, *Phys. Rev. Lett.*, 100, 143902.
51. Holler, M., Zaïr, A., Schapper, F. *et al.* (2009). Ionization effects on spectral signatures of quantum-path interference in high-harmonic generation, *Opt. Express*, 17, 5716–5722.

52. Ganeev, R.A., Hutchison, C., Siegel, T. *et al.* (2011). Quantum path signatures in harmonic spectra from metal plasma, *Phys. Rev. A*, **83**, 063837.

53. Ganeev, R.A., Hutchison, C., Siegel, T. *et al.* (2011). High-order harmonic generation from metal plasmas using 1 kHz laser pulses, *J. Mod. Opt.*, **58**, 819–824.

54. Gaarde, M.B., Salin, F., Constant, E. *et al.* (1999). Spatiotemporal separation of high harmonic radiation into two quantum path components, *Phys. Rev. A*, **59**, 1367–1373.

55. Ganeev, R.A. (2007). High-order harmonic generation in laser plasma: A review of recent achievements, *J. Phys. B: At. Mol. Opt. Phys.*, **40**, R213–R253.

56. Ganeev, R.A. (2012). Harmonic generation in laser-produced plasma containing atoms, ions and clusters: A review, *J. Mod. Opt.*, **59**, 409–439.

57. Ganeev, R.A. (2012). Generation of harmonics of laser radiation in plasmas, *Laser Phys. Lett.*, **9**, 175–194.

Index